T0200406

A Philosophical Guide to Chance

It is a commonplace that scientific inquiry makes extensive use of probabilities, many of which seem to be *objective chances*, describing features of reality that are independent of our minds. Such chances appear to have a number of paradoxical or puzzling features: they appear to be mind-independent facts, but they are intimately connected with rational psychology; they display a temporal asymmetry, but they are supposed to be grounded in physical laws that are time-symmetric; and chances are used to explain and predict frequencies of events, although they cannot be reduced to those frequencies. This book offers an accessible and non-technical introduction to these and other puzzles. Toby Handfield engages with traditional metaphysics and philosophy of science, drawing upon recent work in the foundations of quantum mechanics and thermodynamics to provide a novel account of objective probability that is empirically informed without requiring specialist scientific knowledge.

TOBY HANDFIELD is Senior Lecturer at the Department of Philosophy, Monash University. He is the editor of *Dispositions and Causes* (2009).

A Philosophical Guide to Chance

TOBY HANDFIELD

Monash University

CAMBRIDGE UNIVERSITY PRESS
Cambridge, New York, Melbourne, Madrid, Cape Town,
Singapore, São Paulo, Delhi, Mexico City

Cambridge University Press
The Edinburgh Building, Cambridge CB2 8RU, UK

Published in the United States of America by Cambridge University Press, New York

www.cambridge.org
Information on this title: www.cambridge.org/9781107013780

First published 2012

Printed in the United Kingdom at the University Press, Cambridge

A catalogue record for this publication is available from the British Library

Library of Congress Cataloguing in Publication data
Handfield, Toby.
A philosophical guide to chance : physical probability / Toby Handfield.
    pages   cm
Includes bibliographical references and index.
ISBN 978-1-107-01378-0 (hardback) – ISBN 978-1-107-60735-4 (pbk.)
1. Chance.   2. Probabilities.   I. Title.
BC141.H36   2012
123′.3 – dc23   2011049194

ISBN 978-1-107-01378-0 Hardback
ISBN 978-1-107-60735-4 Paperback

Contents

Preface   [*page* vii]

1 The concept of chance   [1]
    1 An unlucky gamble   [1]
    2 The hallmarks of chance   [2]
    3 Beliefs and probabilities   [8]
    4 Characterising chance   [15]
    5 What makes a fact of chance?   [30]

2 The classical picture: What is the world made of?   [34]
    6 Matter is made of particles   [35]
    7 Particles have properties   [36]
    8 The laws are deterministic   [39]
    9 And that's all there is   [41]
    10 Do the laws heed the direction of time?   [43]

3 Ways the world might be   [47]
    11 A multitude of lists   [47]
    12 Possibilities that differ spatially   [50]
    13 Pushing the limits of possibility   [52]
    14 Creating a space of possibilities   [55]
    15 Possible histories   [59]

4 Possibilities of thought   [62]
    16 Propositions in phase space   [62]
    17 Troublesome thoughts   [63]
    18 Counterfactual possibility   [67]
    19 Macroscopic states   [69]
    20 Phase space and epistemic possibility   [71]

5 Chance in phase space   [72]
    21 The leaking tyre   [72]
    22 Counting possibilities   [73]
    23 Measuring volumes in phase space   [75]

6 Possibilist theories of chance   [78]
    24 Possibilism   [78]
    25 Chances and determinism   [83]

26 Sceptical responses [87]
27 How do we initially grasp the measure over possibilities? [89]
28 How do we make better estimates of chances? [94]
29 Weaker versions of possibilism [96]

7 Actualist theories of chance [104]
   30 Actualist interpretations of chance [104]
   31 Simple actualist proposals [106]
   32 Sophisticated actualist proposals [110]
   33 Can actualism explain the normative role of chance? [119]

8 Anti-realist theories of chance [123]
   34 Varieties of anti-realism [123]
   35 An error theory of chance [124]
   36 Subjectivist interpretations of chance [127]
   37 The subjective psychology of objective chance [131]
   38 Non-cognitivism [142]

9 Chance in quantum physics [146]
   39 The quantum mechanical world [146]
   40 Weird quantum phenomena [147]
   41 The formalism of quantum mechanics [151]
   42 Chance in quantum mechanics [153]

10 Chance in branching worlds [162]
   43 Uncertainty in an Everett universe [162]
   44 Indifference and branches [176]
   45 Bayesian learning about branches [182]
   46 Evaluating the Greaves–Myrvold account [187]

11 Time and evidence [192]
   47 The time asymmetry of chance [192]
   48 Explaining the asymmetry of evidence [197]
   49 Statistical mechanics and the temporal asymmetry of evidence [208]
   50 The roles of evidence, availability, and context [214]

12 Debunking chance [218]
   51 Norms and vindication [219]
   52 The natural history of moral norms [224]
   53 Chance compared to morals [227]
   54 The natural history of chance [231]

References [246]
Index [254]

Preface

When I began writing this book, I believed that I had identified a realist theory of chance which – though not entirely novel – had not been defended as well as it might have been. My book was to have been the definitive presentation and defence of a realist account.

Roughly six years later, I have come to appreciate much better the enormous difficulties facing not only that theory, but all realist accounts of chance, and I find myself in the mildly embarrassing position of writing the preface to a book in which I defend a modest form of anti-realism. In some sense, I now believe, Hume was correct to say that chance has no 'real being' in nature (Hume 1902 [1777]: §8, part I).

During this gradual conversion, becoming better acquainted with the literature, I frequently found the going rather difficult. Much of the literature is very technical, to the point of being inaccessible to many readers, including myself. This is unfortunate. Our best physical theories strongly suggest that chances are a fundamental part of reality. If we are to understand and evaluate these claims, we need to understand philosophical and scientific debates about chance. In consequence, I have written this book, not merely as a vehicle for my own ideas, but also to introduce the philosophy of chance to the broadest possible audience. While I don't pretend that the material is always easy, I expect it should at least be accessible to any tertiary-level reader.

To keep the main line of argument as uncluttered as possible, I have set more technical material and asides which pursue debates of more narrow interest in text boxes. The reader can omit these without fear of losing the main plot. Suggestions for further reading can be found in the footnotes.

I have been fortunate in my friends and colleagues, who have provided invaluable assistance throughout this project.

My greatest thanks go to John Bigelow, Antony Eagle, Patrick Emerton, Lloyd Humberstone, Barry Loewer, and Alastair Wilson, each of whom has read drafts of more than one chapter and provided extremely thoughtful comments and suggestions. I also received some very helpful guidance from an anonymous reader for Cambridge University Press.

Individual chapters benefited from the comments of Stephen Barker, Rachael Briggs, Daniel Cohen, Nina Emery, Steve Gardner, Hilary Greaves, Carl Hoefer, Jenann Ismael, Aidan Lyon, Harry Perlman, Huw Price, and Paul Tappenden. In 2009, a reading group at Monash discussed what was destined to become roughly Chapters 2–7. I am grateful to all the participants, but especially to Alexander Bird, Joshua Luczak, Bryan Paton, and Joel Reicher. Earlier versions of the material were presented to audiences at MIT, Monash, Rutgers, the University of Sydney, Kyung Hee University, and ANU. A sabbatical at MIT in the fall of 2009 was a marvellous environment in which to consolidate my work on the later chapters.

And finally, I'm grateful to many people for stimulating conversations about chance while I have been working on the topic, but most especially to David Albert, John Bigelow, Barry Loewer, Huw Price, Jonathan Schaffer, and Alastair Wilson.

1 | The concept of chance

1 An unlucky gamble

Suppose you were offered the chance to play a simple gambling game, in which you are invited to bet on the outcome of a die-roll. There are only two bets allowed. You can wager that the die will land 6, or you can wager that it will land any of 1, 2, 3, 4, or 5. In either case, if your wager is successful, you will win the same prize: one dollar.

So the bets are:

Die lands 1–5 Pays $1.
Die lands 6 Pays $1.

Assume that you know, moreover, that the die has no significant asymmetry in its construction. It does not have a physical bias to one or more sides.

Which bet ought you to take? Assuming you would prefer more money to less, it is obvious that you ought to take the bet on 1–5, rather than the bet on 6.

Now suppose that you *really do* play this game, and you play it at the same time as a friend. You sensibly choose to bet on 1–5. Your friend, bizarrely, insists that she has a hunch that the die will land 6; so that is the bet she takes. The die lands 6. Your friend wins.

In a sense, this is rather unfair. After all, you took the more sensible bet. You took the bet that you *ought* to have taken. Your friend took the less sensible bet. But it was your friend who won, while you did not. On the other hand, calling this 'unfair' is a touch histrionic. Life is full of chance events, and it is simply the *nature* of chance events that, sometimes, unlikely things happen.

The unfortunate gamble that I just asked you to imagine is a good example of *chance* at work. In this book, I will be attempting to give a general account of what chance is. Before beginning on that task, it will pay to make explicit what I take to be the distinctive features of chance.

2 The hallmarks of chance

Chances are something like 'physical probabilities'

Chances are *chancy*. They are intimately associated with concepts such as likelihood, probability, propensity, and possibility. But not every probability is a chance. What makes chance somewhat different from other probability-like phenomena is that chance is in some sense *physical*.

In emphasising that chances are physical, I mean to highlight two things. The first is that chances are objective. Chances do not depend upon what people believe. At a given time, the chance of a given type of event is the same, no matter who evaluates the chance. Chances are not human creations and in this way they are unlike poems, fictional characters, or lounge suites.

The second point to be highlighted is that chance is the sort of thing that is properly studied by the natural sciences. Physicists, chemists, biologists, and others are all in a good position to give authoritative advice on the chances of various events. (That said, it is often possible to form very well-informed opinions about chances, without any special expertise, special equipment, or lengthy investigation. In particular, most of us are able to make very reliable and rapid inferences from the design of gambling devices about the chance of getting particular outcomes with those devices. Here are some chances that we seem well acquainted with. The chance of: drawing an ace from a well-shuffled pack of cards; a roulette wheel landing on black; getting heads with a toss of a well-made coin; rolling a 6 on a die; etc. In all these cases, there are readily observed physical features of the gambling apparatus which – if the device is used correctly – generate particular chances for the possible outcomes.)

If this all seems too obvious to be worth mentioning, that's all to the good. But just to make sure the point has been hammered home, return to the unlucky gamble in which you bet on the die landing anything but 6. Compare the *chance* that the die would land 1–5 with your and your friend's various *degrees of confidence* that the die would land 1–5. The chance was one and the same for both you and your friend. The chance was the sort of matter on which an applied physicist might have given us authoritative advice, if he or she had the opportunity to make various measurements of the die to check its shape, distribution of mass, and so forth. Neither of these things could be said for your degrees of confidence that the die would land 1–5. First, your degrees of confidence in this outcome were manifestly not the same: you invested most of your confidence in the die landing 1–5. Your friend invested most of hers in it landing 6. So, correspondingly, she must have been less confident than you that it would land 1–5. Second, if we wished to make a more accurate determination of your degrees of confidence, we would not need to consult a physicist. We would not need to study the die. If anything, we would need to examine our respective brains, and we would need the help of a psychologist.

As will be discussed below, degrees of confidence appear to be a species of probability also; but they are very different from chances, in that they are not objective properties of mind-independent processes or events. Rather, they are properties of our minds.

In addition to these ideas about the objectivity of chance and the way we find out about chance, there are two further connotations of the idea that chance is a physical probability. I am less confident that these are strictly part of our concept of chance, so I do not wish to include them as 'hallmarks' proper. Rather I note them here as attractive ideas about chance which might end up being more or less integral to the concept.

The first of these connotations is the thought that chances are fixed by the intrinsic properties of a physical system. Roughly, the idea here is that any two systems that are duplicates – alike in all their intrinsic properties – should also be alike in the sorts of chances that they manifest. Two coins that are physically alike should have the same chance of falling heads. Two atoms of a radioactive element that have the same constitution should have the same chance of decaying. And so on. (There are some difficulties with formulating this requirement precisely, since two qualitatively identical coins, tossed in very different ways, may manifest different chances of landing heads. What we want, in cases like this, is to include the person tossing the coin in the physical system. But then there is a danger that, in order to successfully anticipate all the potential influences from 'outside', we have to treat the *entire universe* as the only relevant system. That would seems to strip this idea of much of its interest.)

The second connotation is the idea that chances can feature in physical – maybe even causal – explanations. For instance: Why did that man get lung cancer? *Because his smoking increased his chances of getting cancer.* Why did that substance emit radiation? *Because each atom of that isotope has a high chance of decaying every few minutes.* These look like good explanations which cite chance.

This is a somewhat controversial idea about chance, because there are quite different views about what explanation involves and what causation involves. Some of these views would suggest that the explanations I have gestured at above are strictly misleading, and that chances are never directly involved in causal explanation.

We ought to believe in accordance with the chances

Your friend, who bet on 6, appeared to be behaving irrationally. Assuming she shared your concern to win the $1 prize, and had no other competing concerns, she chose the suboptimal bet.

What is the *source* of your friend's irrationality? On the information given, we cannot answer this question. There are many different ways in which someone can fail to be rational. Moreover, it is possible that your friend – despite appearances – was in fact rational, because she had access to very different information about the die. (Perhaps a very reliable source had told her that, despite appearing fair, the die is very heavily biased towards 6.)

There is one particular way your friend might have been irrational, however, that is important in understanding what chance is. Suppose your friend agreed with you that the die was fair. So she believed that the die had a ⅙ chance of landing 6. Despite this, her confidence in the proposition 'The die will land 6 on the next roll' was much higher than her confidence that the die would not land 6. In this scenario, your friend has an apparent mismatch between her *belief about the chance* of getting a 6 and her *degree of confidence* that the die will land 6.

We cannot conclusively say whether your friend should have had a different belief about the chance or if she should have had a different degree of confidence. That depends upon the sort of evidence which she has available to her. But we can still conclude that your friend is being irrational because her degree of confidence does not align with her belief about the chances. This is the way chances constrain rational beliefs in general: other things being equal, we ought to apportion our degrees of confidence so as to correspond with what we believe about the chances.

Believing in accordance with the chances is no guarantee of success

You did the rational thing. You bet on 1–5, because that had a much greater chance of paying off than the alternative. But, sadly, you were unlucky, and lost. This is entirely normal for chance events. Even if you adopt sensible beliefs about the chances, that does not mean that all your bets will succeed. Similarly with other aspects of life that are not explicit betting games: if chance is involved, your best-laid plans may still go awry. (The forecast says the chance of rain is low. You leave your umbrella at home . . .)

This gives rise to something of a puzzle, because this feature of chance seems to be in tension with the previous one. *Why* should we believe in accordance with the chances if it does not *guarantee* that we will be better off? Since the most likely outcome is not guaranteed to occur, might we not receive evidence that something particularly unlikely may occur?

There are two replies that are likely to occur to you:

1. *In the long run*, we expect to be more successful if we believe in accordance with the chances than if we adopt any other strategy for forming beliefs. So in some way that is derivative from this claim about the long run, we can draw a conclusion about what is best to believe now, in the single instance.
2. It may be that believing in accordance with the chances is no guarantee, but it is also the case that there is *no better strategy available*. Beggars can't be choosers.

There is some truth in both of these replies, but it will be a task of later chapters to explicate them more clearly, and to assess whether they resolve the puzzle.

Chances of propositions and chances of events

Some philosophers speak of *events* as having chances, others speak of *propositions* having chances. I prefer, in general, to ascribe chances to propositions, since a proposition can refer to an event that does not actually happen, whereas if we speak of events having chances, we need always to qualify that not only actual, but merely possible events can have chances.

It does not seem likely that anything of great import turns on this distinction. There is generally a straightforward translation from a claim that an event of type E has chance x to the claim that ascribes chance x to a *proposition*, which states that an event of type E will occur.

Something interesting does turn on whether we ascribe chances to *propositions* or to *sentences*, however. The same proposition can be expressed by different sentences, and it has been a matter of ongoing interest to philosophers that two sentences which express the same proposition can have very different cognitive significance. For example:

(1) Jack the Ripper killed at least five women in 1888

is a relatively uninformative sentence. But if Jack the Ripper is in fact the name of a man otherwise known as Tom Smith, then the following sentence is also true:

(2) Tom Smith killed at least five women in 1888.

Arguably, these two sentences express the same proposition. The names 'Jack the Ripper' and 'Tom Smith' both denote the same man and the same property is ascribed in both sentences. On commonly held views about propositions, these facts suffice to determine that the propositions are the same.

But, clearly, although they might express the same proposition, the cognitive significance of these sentences is very different. No well-informed person would be surprised to hear the first, but the second would be very big news.

How does this bear on matters of chance? Take another example (drawn from Hawthorne and Lasonen-Aarnio 2009), in which we create an artificial name for a lottery ticket, in a lottery that is yet to be drawn. We declare: 'The ticket that will win this lottery is named "Lucky".' With this name to hand, I can now confidently assert that:

(3) Lucky will win the lottery.

This sentence is guaranteed to be true, given the special way I created the name 'Lucky'. So I should have the highest possible degree of confidence that this sentence is true. If there are chances that attach to *sentences*, this sentence has a chance equal to one.

Now consider, for each of the 10,000 tickets in the lottery, sentences of the form:

(*) Ticket number N will win the lottery.

Assuming the lottery is fair, presumably each of these sentences has a chance of 1 in 10,000. But one of these sentences expresses the very same proposition as (3). So we have two sentences that express the same proposition, but they ascribe different chances. This gives rise to some awkward questions. Believing that a proposition is true while also believing that it is false is to be caught in a contradiction. Similarly, it seems contradictory to ascribe to a proposition two different chances. But if chances attach to sentences rather than propositions, it seems that we will have to ascribe different chances to the same proposition.

If, on the other hand, we ascribe chances to propositions rather than to sentences, we need to choose what chance we ascribe to the proposition expressed by (3). Presumably, this proposition cannot have a chance of one, simply because we have created a name like 'Lucky'. Otherwise, all chance claims would risk being trivialised as soon as we had created a name of this sort. So the chance of (3) must be only 1 in 10,000, even though we are absolutely certain that the sentence is true! Now we have a problem formulating our requirement that chances should accord with our degrees of belief. For in this case, our degree of belief in the *sentence* (3) should *not* match the chance of the corresponding *proposition*.

I won't go further here into precisely how this should be handled. I will simply work on the idea that our degrees of confidence should – other things being equal – match the sorts of chances that attach to propositions. The complexities related to our degree of confidence in sentences that involve names like 'Lucky' and 'Jack the Ripper' I leave for another occasion.

Chances change over time

A less obvious feature of chance is that the chances of a proposition can change over time. To bring this out, consider the chance that the die lands 6. Before the toss, the chance of this was something like $\frac{1}{6}$. Accordingly, you were not certain that the die would land 6. But now that it has landed 6, your degree of belief has changed: you are certain that it landed 6. Is this because you now believe contrary to the chances? Surely not: rather, it is because the chance has *changed*, from $\frac{1}{6}$ to 1.

More generally, there seems to be a pattern in which propositions about future events can have chances that are greater than zero and less than one – what I'll call non-trivial chances – but past events frequently have only trivial chances: zero or one.

That is not to say that all propositions about the past have trivial chances. Consider the claim that *Napoleon had, in his entire life, an even number of meals*. It is very hard to be certain of a claim like this. It is hard to envisage how it could ever be settled. Consequently, depending somewhat upon your views about chance, you might think that this proposition – and other propositions about the past – can have a non-trivial chance also.

For readers who are unconvinced that Napoleon's culinary history can be a matter of chance, there are other examples which, though more contrived, are perhaps more intuitively compelling. For instance, suppose that you have a time machine which can take you back to the moment of Napoleon's birth. As you begin your time-travel journey, you toss a coin a long way in the air. By the time the coin lands, both you and the coin will have arrived at that earlier time. The proposition 'This coin lands heads at the moment of Napoleon's birth' might appear to be a matter of non-trivial chance, even though the event occurs in the past.

All of this invites the inquiry: what *makes* the chances change over time? A satisfactory account of chance should give some idea of how or why this occurs.

Changing chances over time

Note that it is still true to say that the chance the die would land 6 *was*, before the die was rolled, $\frac{1}{6}$. So in what sense has the chance changed?

We can best get the distinction needed clear by using some formal notation. I'll write the proposition in question, such as 'Napoleon consumed an even number of meals in his entire life', as P. For the chance that P at a given time t, I'll write: Chance-at-$t(P)$. So the claim that chances can vary over time is simply the claim that it is not always the case that, for all times, t_1 and t_2, Chance-at-$t_1(P)$ = Chance-at-$t_2(P)$.

The claim that the chance was $\frac{1}{6}$, then, is simply the claim that there is a time t_0, earlier than now, such that Chance-at-$t_0(P) = \frac{1}{6}$. That chance claim is true now – at t_1 – even though Chance-at-$t_1(P) = 1$.

It is true at t_1 that Chance-at-$t_1(P) = 1$.
It is true at t_1 that Chance-at-$t_0(P) = \frac{1}{6}$.

The potential for confusion arises because there are two times involved: the time at which we evaluate the claim and the time involved in the chance itself.

---

3 Beliefs and probabilities

Degrees of belief

Before attempting to characterise chance, it is necessary to introduce an important technical notion: the idea of 'degree of belief' (also referred to as 'credence'). In ordinary practice, we simply talk about believing or disbelieving. Sometimes we might admit talk of 'partial' belief. But it is quite unfamiliar to think that we might assign these partial beliefs numerical grades.[1]

Even if we do not endorse describing our conscious thought in these terms, there are a number of convergent reasons to think that we might possess a mental state that is at least importantly similar to belief, yet which comes in numerical degrees.

One reason is that there are scenarios where we have *limited*, but *precise*, information about what is going to happen in the future, and if we are to use that limited information in the best way possible, we will need to 'divide our minds' between different possibilities, in some sense. Consider the mental state which you might have adopted with regard to the gamble introduced at the beginning of this chapter. Suppose you played that gambling game six times, and you were reliably informed that the die would land 6 on only one occasion. If, for each trial, you could adopt only one fixed state of mind with respect to the propositions, 'the die lands 6 on trial 1', '. . . on trial 2', and so on, what would the best such state of mind be? Whatever state of belief that is, we might be able effectively to *define* degree of belief ⅙ to be the attitude that is best to have in a scenario such as this, where the event in question happens once in six trials.[2]

A second way of getting at the concept of degree of belief is to consider how you would value various possible gambles. This was the sort of strategy I was using in the story of the gamble at the beginning of the chapter. The fact that you preferred the gamble on 1–5 over the gamble on 6 was evidence that you had a higher degree of belief that the die would land 1–5.

In order to develop this strategy so as to determine precise numerical degrees of belief, we would need to vary the relative value of the prizes for the two gambles. If a more valuable prize is attached to a gamble,

---

1 Psychological evidence confirms that people prefer to use qualitative terms to describe their own psychological states, rather than numerical terms, even when allowed some vagueness for the numerical expressions (Budescu and Wallsten 1995: 297–8), although, interestingly, the same studies suggest that most people prefer to *receive* probabilistic information *from others* in numerical form, rather than in qualitative terms.

Richard Holton (2008: 35–40) makes a helpful philosophical proposal to characterise the ideas of belief and partial belief, as opposed to the concept of credence. Holton suggests that all-out belief involves resolving the unmanageable amount of information involved in our credal state into a single possibility which one takes as a 'live possibility': a basis for deliberation. So one all-out believes that P if and only if one takes P to be a live possibility, and does not take not-P as a live possibility. Partial belief is a relaxation of all-out belief. One partially believes P if (and only if) one takes P as a live possibility, but also takes not-P as a live possibility.

2 F. P. Ramsey used this sort of characterisation when he wrote that 'belief of degree $^m/_n$ is the sort of belief which leads to the action which would be best if repeated n times in m of which the proposition is true' (Ramsey 1931: 188). See also Galavotti (2005: 202).

then it becomes more attractive to take it. When the two gambles seem equally desirable, it is possible to translate the ratio of the prize-values into information about relative degrees of belief. For instance – speaking for myself – I value $5 roughly five times more than I value $1. And, moreover, if the prize for betting on 6 was $5 and the prize for betting 1–5 was $1, then I would be indifferent between the two gambles. This reflects my confidence that it is five times less likely that the die will land 6 than it is that it will land 1–5. Accordingly, my credence that the die will land 1–5 is ⅚, and my credence that the die will land 6 is ⅙.

> To turn the different prize-values into a degree of belief, you use the formula:
>
> $$\text{Degree of belief(Gamble 1 wins)} = \frac{\text{Value(Prize 2)}}{\text{Value(Prize 1)} + \text{Value(Prize 2)}}$$
>
> This idea has been defended by a number of writers. I am largely following D. H. Mellor (1971: chap. 2) in my presentation of the idea.

Unfortunately, this rather indirect approach to explaining a degree of belief suffers from the problem that it involves psychological assumptions that may be utterly unrealistic. Many perfectly rational people object to gambling on a variety of grounds. Asking them to contemplate what they would think about being offered possible gambles might return the disappointing result that they would reject all such gambles. But simply being inclined to reject such gambling games surely does not suffice to show that these individuals have no credences.[3]

A further problem is that it supposes that our valuations of prizes are mathematically well behaved, so that we can, for instance, meaningfully say that a person values one thing 'five times more' than she values something else. It is not clear that this is so.

Without going into the full details of debates in the theory of value, note that while it might seem normal for me to be indifferent between the two gambles on the die when the prizes are $1 and $5, it would be surprising

---

3 Ramsey addressed this point briefly:
> Whenever we go to the station we are betting that a train will really run, and if we had not a sufficient degree of belief in this we should decline the bet and stay at home. The options God gives us are always conditional on our guessing whether a certain proposition is true.   (Ramsey 1931: 183)

Though Ramsey is quite correct that our choice of action is almost invariably based upon 'guesses' as to what will happen as a result of our choice, this is probably not an entirely adequate reply to the worry. The concern is that our degrees of belief may not be as *tightly linked* to our betting dispositions as is required by the Ramseyan method of identifying credences. So some degree of idealisation remains inevitable.

if I maintained that indifference when the prizes became $1,000,000 and $5,000,000. I would much rather have a large chance at the smaller prize than have a small chance at the larger one. Whether this can be explained without disrupting the elegant mathematical simplicity of this approach to defining credences is a matter of ongoing controversy.[4]

So if the method requires entertaining a largely hypothetical psychology, and perhaps other idealising assumptions to ensure that our valuations are mathematically well behaved, then the method appears to show only that a hypothetical, gambling-tolerant person, loosely based upon you, has numerical degrees of belief. But it does not clearly show that *you* have them.

However, the third line of evidence goes some way to answering this concern that degrees of belief might be merely hypothetical constructs. There is emerging psychological evidence that humans are capable of excellent probabilistic reasoning about some matters – and such probabilistic reasoning requires that there be some mental state that is akin to a degree of belief.[5] What is notable about these types of reasoning, however, is that they are frequently *unconscious*. For instance, psychologists have used probabilistic models successfully to understand the processes by which we infer causal relationships from perception, such as when we see one billiard ball hit another and infer that the motion of the second was caused by the collision with the first. Other examples include the processing of visual information in the act of perception and the acquisition of language by infants. None of these processes typically involves conscious contemplation of probabilistic information. So the best evidence for 'degrees of belief' seems to be at an unconscious level of mental life, which goes some way towards explaining why there may be resistance to the idea as a species of *belief*, given that belief is stereotypically a conscious attitude.[6]

---

4 Leonard Savage (1954: 101–3) gives a straightforward presentation of the basic problem of risk aversion – though Savage tries to argue that the phenomenon is irrational. See Allais and Hagen (1979) for some early, relatively technical literature on risk-aversion and how it can be accommodated by probabilistic decision theory. Much more recent and more empirically informed work on the psychology of risky decisions is collected in Kahneman and Tversky (2000).

5 See Chater, Tenenbaum, and Yuille (2006) for an overview of these developments in cognitive psychology.

6 Note that the strategy I have employed to explain degrees of belief has not given an *analysis*. Rather, I have indicated some phenomena that seem to be associated with degrees of belief, such as hypothetical roles such a mental state might play, and apparent manifestation of that state in unconscious inference. None of this, however, amounts to a proper analysis or definition of the concept. Eriksson and Hájek 2007 argue that the concept is in fact not capable of being analysed, and must be taken as a primitive. But because there are a number of important

The logic of degrees of belief

Ordinary belief is subject to certain logical requirements. Someone who believes both that 'Nixon was president of the USA' and also that 'Nixon was never president of the USA' is prone to get into practical trouble working out how to answer quiz questions about US presidents. But he is not merely liable to have practical difficulties, he is also *inconsistent*. This is a criticism that can be made of his psychology on purely logical grounds.

Are credences subject to any logical requirements in the same way as ordinary belief? Here are some examples of credences that appear to be incoherent:

- Jill's credence that her cat is hungry is 0.9. Her credence that her cat is not hungry is 0.2.
- Kevin's credence that this coin will land heads is 0.5. His credence that it will land tails is 0.5. But his credence that it will land *either* heads or tails is only 0.9. (And Kevin does not think that the coin can land *both* heads and tails at the same time! He has credence zero in that proposition.)

In order to avoid perversity such as Jill's or Kevin's, credences need to comply with a set of mathematical axioms which define the *probability calculus*. Mathematically speaking, a probability is any mathematical function which obeys these rules. Since credences must obey these rules, as the above examples are intended to show, we can reasonably say that *credences are themselves probabilities*.

But of course appeal to specific examples is not a decisive way of showing that our credences must, *always* and *everywhere*, obey the axioms of probability. A number of writers have tried to give a rigorous proof that an agent must conform his or her credences to the axioms. These 'Dutch Book' arguments purport to show that, for anyone who has degrees of belief, if those beliefs do not obey the axioms of the probability calculus, then they will be disposed to accept a series of bets on which they are doomed to lose.[7]

To refer back to the examples, suppose we are unsure which of Kevin's credences about the coin is most unsound. Perhaps it is a very strange coin that frequently lands on its edge. So Kevin could be correct in thinking that the probability that it lands either heads or tails is only 0.9. But, in that case,

---

constraints on how degrees of belief function, they are optimistic that it is still a relatively well-understood concept.

7 The manner of argument goes back to Ramsey (1931). See Hájek (2005) for a valuable recent discussion and refinement of the argument.

he is wrong to have such high credences in heads and tails separately. The 'Dutch bookie' can exploit this, without needing to work out whether or not the coin is indeed bizarrely liable to land on its edge. He could offer Kevin, for instance, the following bets that pay out $10 each:

Bet on heads   Kevin will be willing to pay up to $5 for this wager.
Bet on tails   Kevin will be willing to pay up to $5 for this wager.

So suppose Kevin pays out a total of $10 in purchasing these two bets. But then the bookie also asks if Kevin would be willing to sell a bet to the bookie.

Bet on heads or tails   Kevin will be willing to sell a bet that pays $10 if the coin lands either heads or tails, at a minimum price of $9.

Suppose Kevin now receives $9 from the bookie in selling this bet.

Now, Kevin has paid out $1 more than he has received. So he has lost from the transactions; and no matter what happens with the coin, this won't change. If the coin lands heads, Kevin collects his prize of $10, but he also pays out $10 to the bookie. If the coin lands tails, the same applies. And if the coin lands on its edge, none of the bets win, so no one pays out.

The attempted proof generalises this strategy, showing how the bookie can construct packages of bets to buy and sell, which together will trap the wayward bettor into a certain loss. If such a proof succeeds, it would certainly show that there is something deeply problematic about defying the probability calculus.[8] But as is evident from the example of Kevin, these arguments tend to inherit many of the problems mentioned above, related to requiring that degrees of belief translate straightforwardly into dispositions to accept and reject certain gambles. Couldn't Kevin have those bizarre credences, but be protected from the Dutch bookie by his hatred of gambling? Consequently, it seems unlikely that a truly watertight proof of this sort can be given.[9]

---

8 Strictly speaking, it is not enough merely to show that violating the axioms of probability leads to vulnerability to a Dutch Book. We also need to show that conforming to the axioms of probability leads to *invulnerability* to a Dutch Book. Such a proof is known as a converse Dutch Book result (Kemeny 1955; Lehman 1955).

9 Another limitation of Dutch Book results, pointed out by Kyburg (1978), is that it is possible for an agent to recognise a Dutch Book as involving a sure loss, without using probabilistic concepts at all. It is merely a matter of deductive logic that any agent who accepts a combination of bets like that will be doomed to lose. One simply needs to enumerate the possible outcomes, and the cumulative result of the combined bets, and one can see that a sure loss is involved. Hence Kyburg suggests that all an agent needs to be protected from a Dutch Book is an aversion to sure losses, and the ability to engage in non-probabilistic reasoning.

In a brief remark, F. P. Ramsey neatly hit upon what seems to be the underlying problem with adopting degrees of belief that violate the probability calculus: 'If anyone's mental condition violated these laws, his choice would depend upon the precise form in which the options were offered him, which would be absurd' (1931: 182). But I fear that, as stated, this is still not sufficient to prove the case. As noted in the box above, 'Chances of propositions and chances of events', if you offer me a bet on whether the ticket 'Lucky' will win the lottery, where this is a special name given to the ticket that will win the lottery, then I will be very keen to accept this bet. It is presumably rational for me to have a very different attitude to a bet on whether ticket number 3,084 will win the lottery – though it may well be that these are the very same options, differently described. Without further explanation, Ramsey's suggestion is at odds with these attitudes.

Despite the difficulty in constructing a rigorous proof of this point, it remains extremely plausible that there is usually something seriously awry with someone's credences if they do not obey the probability calculus. No doubt none of us do perfectly conform, in all of our degrees of belief, to those rules. Nobody is perfectly rational. But the ideal remains an attractive one.

The axioms of probability

The axioms of probability calculus are best known through their formulation by Kolmogorov (1956: §1). For a more thorough introduction to formal aspects of probability theory, I recommend the excellent primer at the beginning of Eagle (2011).

Probability is a function that maps each proposition onto a single real number, between zero and one.

Adopting 'prob(·)' as our way of representing a probability function, the expression:

$$\text{prob}(Q) = x$$

means that the probability of proposition Q is x.

It is not clear to me, however, that this is in fact a serious defect. Dutch Book results show that *if* subjective states of mind such as credences can be linked to betting behaviour in the way supposed by the Dutch Book results, then there is an independent way – outside of probability theory – of showing that there is something problematic about those states of mind not conforming to probability axioms. If we could *only* show that there is a problem with incoherent credences by appeal to *probabilistic* facts, then that would cast doubt on the significance of the entire result. It is true that we might be able to avoid Dutch Books by avoiding credences. But *if* we want to use degrees of belief, and to tie those degrees of belief to betting behaviour in some way, then this looks like a good reason to use the probability calculus to regulate those beliefs.

We then have three axioms:

A1. If P is logically equivalent to Q, $\mathrm{prob}(P) = \mathrm{prob}(Q)$.
A2. If P is inconsistent with Q, then $\mathrm{prob}(P \text{ or } Q) = \mathrm{prob}(P) + \mathrm{prob}(Q)$.
A3. If P is a priori true, then $\mathrm{prob}(P) = 1$.

Almost all of the standard things we might wish to say about the basic logic of probability follows from these axioms. What they do not give us, however, is the notion of a *conditional probability*. Intuitively, a conditional probability is the idea involved in forms of speech such as: 'I'm not likely to win the prize if Jones enters'; or 'Given last year's winner was not very dexterous, I have a good chance.' In statements like these, the probability statement is not unconditional, but made conditionally upon a particular proposition, which may or may not be true.

I will write the conditional probability of P, given Q as:

$$\mathrm{prob}(P, \text{ given } Q).$$

Kolmogorov suggested that conditional probabilities could be defined in terms of unconditional ones, as follows:

$$\text{C1.} \quad \mathrm{prob}(P, \text{given } Q) = \frac{\mathrm{prob}(P \text{ and } Q)}{\mathrm{prob}(Q)}$$

But this has turned out to be controversial, for a number of reasons. The easiest of these to observe is that there are sometimes conditional probabilities, $\mathrm{prob}(P, \text{given } Q)$, even when $\mathrm{prob}(Q) = 0$. But according to (C1), the conditional probability is undefined when the probability of Q is zero. (See Hájek 2003b for discussion.) Alternative axiomatisations allow us to overcome this problem by taking conditional probability to be the primitive notion, and unconditional probabilities are analysed in terms of conditional ones.

4 Characterising chance

With the concept of a degree of belief now in hand, we can proceed to characterise chance itself. It is helpful to conceive of chance as something like an 'advisor'.

Chance as an advisor

We can precisely characterise different *advice functions*. These functions take a proposition as input and recommend a degree of belief as output. So an advice function is much like a fortune-teller. You ask 'Will I live a long and healthy life?', and the function will output a recommended credence in the proposition that you will live a long and healthy life. We might write this as:

$$\text{Advice(I live a long and healthy life)} = x$$

where x is the recommended degree of belief that is output by the advice function.

Advice functions are merely mathematical objects. Such a function can be thought of as a long list (infinitely long) of input–output pairs. There is no doubt that such functions 'exist' in whatever sense mathematical objects exist. The important question is: which ones are of interest to us? Is there one advice function that corresponds to the special role of chance?

There are some functions which would be good representations of your subjective opinion about various matters. You could think of this as the advice *you* would give, were you asked for it, and wanted to be sincerely helpful. For instance, you might be much more confident that the sun will rise tomorrow than you are that you will catch chickenpox next year. So an advice function which returned values like:

$$\text{Advice(Sun rises tomorrow)} = 0.9$$
$$\text{Advice(I catch chickenpox next year)} = 0.1$$

might be eligible to represent your beliefs. Functions which make the reverse recommendations (i.e. recommending a high degree of belief in catching chickenpox, and a low degree of belief in the sun rising) are ineligible to represent your beliefs, because they do not capture the right relations between your beliefs.

It is unlikely, however, that just *one* function will be eligible to represent your beliefs. This is because it is implausible that your psychological state determines, for each proposition, a *unique* number which accurately represents your degree of belief in that proposition. There is too much sloppiness and indeterminacy in our mental states to allow such a precise assignment. But if objective chance is thought of in terms of such an advice function, then it seems natural to think that the world must be 'nominating' one advice function as the uniquely preferred one. The advice given by this function is the degree of belief which is, given the way that the world is, the best possible advice for what to believe. This is a very rough, first approximation to what we mean by *objective chance*.[10]

So our first approximation is:

Chance-1 The chance that P (at a time t, in a world w) is the degree of belief in P that is recommended by the best possible advice function (for w, at t).

This, however, turns out not to characterise the notion of chance. Instead, it has characterised another sort of advice function which looks even better than chance in terms of the quality of advice it gives. This is simply the *omniscient function*. This function will take propositions about any subject matter, and returns the value one if they are true and zero if they are false. If we set our degrees of belief accordingly, we will believe all the truths and reject all the falsehoods.

The omniscient advice function certainly exists, in whatever sense mathematical objects are said to exist. And it is clear that it gives the best possible advice. Of course, we don't know *which* function, among the myriad of possible advice functions, really is *the* omniscient function. But still, once we have noted the existence of such a function, it is important to try to get clear on why we are interested in a different advice function, such as the envisaged chance function, which seems to give *less useful* advice. At the same time, we hope to obtain a better understanding of the chance function: what role does it play that properly distinguishes it from the omniscient function?

One thought that often occurs to people is to say that chance is something like an optimal guide to belief, given only *partial* information about the world. In particular, many people have thought that chance should be characterised as something like an optimal advice function, given only information about the *past*:

Chance-2 The chance that P (at a time t, in a world w) is the degree of belief in P that is recommended by the best possible advice function, given only information about what has happened in w, up to t.

This is a bit unclear, however, as stated. Who is it that is 'given information'? Advice functions, as I characterised them, are simply given a proposition P, and return a recommended credence. It is doubtful how to make sense

---

10 It is probably too demanding to insist that there be only *one* advice function that plays this role. Some chances might be indeterminate, in which case different advice functions that give different advice on the indeterminate chances might all be equally good candidates. So, strictly speaking, we are seeking a *class* of advice functions that are all equally good candidates to be 'the' chance function of the world.

of giving an advice function its regular input, P, as well as some additional information about the past, and getting a different result. Moreover, it remains unclear why Chance-2 will do any better than Chance-1: both seem to pick out the omniscient function. If the *best* thing to believe about P, without qualification, is the truth about P, then why would anything *less than the truth* be best, given *additional* information?

One influential way of characterising chance that avoids some of this confusion is David Lewis's 'Principal Principle'. This principle makes explicit the idea that the advice given by chance is in some sense optimal, without trivialising chance, by characterising it as a sort of limited epistemic 'trump'. That is, given the advice provided by the chance function, this advice is *not* the best possible advice. But it is good enough advice to trump any other 'admissible' evidence. Whatever other admissible evidence you have, you should nonetheless make your degree of belief conform to the advice given by the chance function.[11]

Using 'prob$_o$' to represent a probability function that corresponds to rational degrees of belief, prior to having received any evidence, 'E_t' for any admissible evidence at a time t, and 'ch$_t(\cdot)$' for the function giving the chance of the proposition in the parentheses at t, we can formalise the Principal Principle as:

$$PP : \mathrm{prob}_o(P, \text{ given } E_t \text{ and } \mathrm{ch}_t(P) = x) = x$$

Expressed informally, any rational initial credence function will return, conditional on the chance of P, and any other admissible evidence E_t, a credence in P that equals the chance.

Lewis's characterisation, however, is not adequate until we provide an explicit understanding of what makes evidence 'admissible'. For instance, if *everything* is admissible, then the chance function will just be the same as the omniscient function. The only advice that is going to be robust in the face of the totality of facts is the outright truth. But ordinarily we think of chance as playing a separate role from the omniscient function, so we must have some principled reason to restrict what is admissible. As I noted above, many have thought that chance is something like an objectively optimal

---

11 Lewis introduced his 'Principal Principle' in Lewis (1980: 87); see pp. 92–6 for his full characterisation of admissibility. He later refined his version of the Principal Principle in response to a conflict between the original principle and his other metaphysical commitments (1994). There has been a significant literature on the topic since. See, e.g. Briggs (2009a); Hall (1994, 2004); Hoefer (1997); Schaffer (2003b); Strevens (1995); Thau (1994).

guide to belief, given we only have information about *the present and the past*. So one way of characterising admissibility is simply that all and only information about the past is admissible. Information about the future is inadmissible. A common variation on this idea that makes chance an even stronger trump is to think that information about the laws of nature should be admissible also.

This leads us to something like the following:

Chance-3 The chance that P (in w at t) is the degree of belief recommended by an advice function, whose advice any rational agent would accept, conditional on knowing this advice and anything else about the past and the laws of nature.

Because Chance-3 is silent on what a rational agent would accept, if she knows information *other* than just the advice given by the chance function and information about the past, it remains open that the chance function can be 'overtrumped' by better information. Hence chance can be different from the *omniscient* function, which cannot be outranked. But note that the lesser status of chance is not guaranteed by Lewis's characterisation. Chance-3 no longer uses the 'best' formulation that was used in the first two characterisations – so it no longer forces the trivialisation of chance upon us – but it still leaves it open that the chance function could simply be trivialised as the omniscient function.

We have run into one serious problem with our attempt to characterise chance: it is hard to explain what chance is, without running the risk that chance is trivially equivalent to truth or falsity. I will return to this problem below. But first, I wish to register a disagreement with Lewis's idea that chance is advice that can be outranked by information from the future.

We do better, I claim, to identify admissible information with any information that is *available*. Almost always, the available information is the *same* as information about the past. We have excellent available information about the temperature yesterday, whereas we do not have anything more than cunning attempts to predict the temperature tomorrow. We have available to us excellent information about the size and shape of the pyramids one thousand years ago, but practically no idea what condition they will be in one thousand years hence. Note that by *available* information, I do not mean information that I actually *have*. Nor do I mean information that any person *actually possesses*. Rather, I mean information that *can be possessed*. There are weather instruments which have recorded the temperature yesterday. There are historical accounts and archaeological traces which could inform us about the condition of the pyramids one thousand years ago, even if no one has read those accounts or inspected those records. But there is no similarly available evidence about the future.

So my characterisation of admissible evidence as the available evidence coincides quite closely with Lewis's characterisation as information about the past and the laws. But in one sense, availability is a *more stringent* constraint, because there are limits to what we can know about the past. I tacitly appealed to this idea in the Napoleon example above, where I suggested that there might be a matter of chance regarding the proposition that Napoleon ate, in total, an even number of meals during his life. On Lewis's view, there could be no non-trivial chances regarding past matters like this (Lewis 1980: 92–4).

In another sense, restricting admissible evidence to what is available may be *less* stringent than the traditional restriction to information about the past and the laws. It might be possible for information from the *future* to be available. This might be the case if some sort of time travel were physically possible. Suppose I received a message from the future that in next week's lotto draw the first three numbers drawn will match the numbers on my ticket. 'I now have a much higher chance of winning', I say. Those who think that only past information is admissible would say that I have erred: the chance itself is *unchanged* – but having received inadmissible evidence, I am no longer rationally required to believe in accordance with the chances. Though I place no great weight on this point, I think that common usage favours my approach here.

Here is another example of common usage that appears to favour my approach. Lewis believes that, if the laws are deterministic, nothing is left to chance. This is because, if the laws are deterministic, then full information about the past, conjoined with the laws, entails all future truths. So the only advice about the future that is robust enough not to be trumped by information about the past and the laws is the advice given by the omniscient function. So there are no non-trivial chances.

But as things actually stand, it is quite possible that the fundamental natural laws *really are* deterministic. Although it is widely believed that quantum mechanics is an indeterministic theory, it is currently a matter of live debate among physicists which version of quantum mechanics is correct, and some of the options are indeed deterministic (we will review these options in Chapter 9). So there is – according to Lewis – a live threat to the existence of non-trivial chances.

Nonetheless, it seems obviously true to say that the chance that this coin lands heads, if I give it a decent toss, is ½. The *possibility* that the laws of physics might turn out to be deterministic seems completely irrelevant to this claim. If someone were to complain that I don't *really know* this chancy fact, because the laws might be deterministic, we would be within our rights to regard this person as eccentric in their understanding of chance. My proposal can allow that there are chances in a deterministic world, because although there is a lot of information available about the past, not *all* information about the past is available. So we cannot derive all truths about the future from the available evidence about the past conjoined with deterministic laws.

Again, this is a small point in favour of my proposal. But again, I don't think we should place a great deal of weight on this sort of evidence. For now, I will provisionally assume the desirability of using the notion of availability in our notion of chance. I return to the case in favour of it a little later.

We now have two modifications to incorporate into our characterisation of chance, to improve over Chance-3. First, we wish to get a better grip on the circumstances in which non-trivial chances will exist. Chance-3 leaves it open that, in every world, the only chances are trivial. Second, we wish to use the notion of available evidence, rather than the notion of evidence about the past and the laws, in order to characterise the 'trumping power' of chance.

Here is a rough thought as to the important role of the chance concept in our interactions with each other. When I say, 'There is a high chance of rain', I am not merely suggesting that I have evidence which recommends a high credence in rain. Rather, I am claiming that *there is no identifiably better opinion to be had* on the matter. This is a claim that will, if successful, prevent you from trying to gather further evidence on the matter – at least for now. This seems to me to capture something very important about the role of chance talk.

So in this vein, I suggest the following characterisation of chance:

Chance-4 The chance that P (at a time t, in a world w) is the degree of belief in P that is recommended by the best identifiable advice function, given only information that is available at t.

This approach clarifies what was unclear in Chance-2, when we wanted to say that chance was the best advice, 'given certain information'. What we mean by that is that it is the best *advisor we can choose*, given certain information. We have also, in co-opting and revising Lewis's idea about admissibility, identified the relevant information as that which is available.

I doubt that this characterisation will suffice as a definition or as an analysis. I do not propose it for that purpose.[12] Rather, I merely claim that Chance-4 is a helpful way of characterising the central topic of this book, such that we give due emphasis to some of its most philosophically important features.[13] Below, I expand on some of these features.

The central role of chance is normative

This characterisation takes the second hallmark of chance – the role in prescribing epistemically appropriate beliefs – to be the central one. The upshot is that chance is, as I understand it, a normative concept. Claims about chance are primarily claims about how things *ought to be*, rather than claims about how things are. This is not particularly controversial in recent philosophical discussions, but it is not always given as much emphasis as is warranted.

Chance clearly plays a role in scientific theorising and in scientific descriptions of the world. But it conflicts with various traditional views in philosophy to think that science uses normative concepts to describe the world.[14] Moreover, science purports to describe the world in entirely naturalistic terms. But ethical concepts, and by extension other normative concepts, are notoriously difficult to analyse in naturalistic terms. So by emphasising the normativity of chance, I am (deliberately) emphasising sources of likely

---

12 One reason to think that Chance-4 is inadequate as an analysis is that it drops all mention of the *physicality* of chance. Suppose that there is *no* available evidence about some matter P or not-P. It might be that the best identifiable advice about this matter is simply to divide your credence equally between the possibilities. But this is arguably insufficient for P to have a chance of 0.5.

13 For readers interested in further debates about how chance can be understood as an advice function, and the related debate about Lewis's Principal Principle, I recommend Hall (2004) and Joyce (2007) as excellent entry points to the literature.

14 The stereotypical example of the view that science is non-normative is logical positivism. See, e.g. Ayer (1952: chap. 6), where Ayer asserts that ethical concepts are 'pseudo-concepts' (p. 107) and consequently ethical statements are not 'factual', and could not be part of a science (p. 112). Ayer and other positivists had much the same attitude to all normative language.

Robert Black (1998: 384) nicely articulates the idea that chances are normative when he writes:

> Credence-makers look metaphysically 'queer' in rather the way in which John Mackie thought objective values queer – they build a strange bridge between what is 'out there' in the world (the propensity [or chance]) and something from a different conceptual realm (reasonableness of confidence in an occurrence). It's as though the realist [about chance] were claiming that in order to be rational one has to regard oneself as inhabiting an Alice-in-Wonderland world whose objects were decked out with labels saying 'Believe this . . . !', and furthermore, that the labels are not mere decoration, but are indeed objectively to-be-believed.

trouble. How can a normative concept be accommodated in an entirely naturalistic account of the world? Why does science need an inherently normative concept to describe the world? No one would define mass as 'the best way to explain why bodies resist acceleration'! These issues cannot be addressed immediately, but are important to bear in mind as we proceed.

The idea of 'available evidence'

The concept of 'availability' is somewhat slippery. Here is a scenario that illustrates some of this slipperiness. Suppose that John has some symptoms that indicate there is a higher than normal risk that he has prostate cancer. He sees his doctor, who advises him that he should not be too worried, as there is only a 10 per cent chance, given his age and symptoms, that he has a malignant tumour. The doctor recommends a biopsy, however, to make sure.

John has the biopsy performed, and the pathology lab conducts a test on the tissue. The test results have arrived in the doctor's office, and John is about to leave home to see the doctor and find out the results. But just as he is about to leave, a concerned friend rings up, having just heard the news that John may be unwell. The friend asks, 'What is the chance that you have a malignancy?' John replies, 'Don't worry. It's only 10 per cent.'

Such a reply seems entirely appropriate. (Though that is not to insist that it is the *only* legitimate way of responding.)

Now suppose that John is in the waiting room of the doctor's and he receives a second call from a different friend. This friend asks the same question as the first. John might now say: 'I don't know the chance yet. Ask me in ten minutes and I'll tell you.'[15]

This reply, too, seems appropriate. One factor that might explain the appropriateness of these two different replies is that, when John is in the waiting room, a crucial piece of evidence – the result of the pathology test – has become 'available' in some sense. Even though John still does not have it, it is clear that he *can get it*. So it no longer seems appropriate to rely upon the doctor's earlier estimate of the chance, which was based upon a much more limited stock of available evidence.

The suggestion, then, is that the available evidence can change, and that this can explain changes in the chances. In this particular instance, no additional tests were performed between the first phone call and

---

15 This example is adapted from DeRose (1991: 582–6).

the second call. Rather, the salient thing that changed was John's proximity to obtaining the test result. This suggests that what is *available* depends, in part, on the circumstances of the person using that concept. At home, John's circumstances were such that the new test result was not available. But when John is in the doctor's office, the new test result is available.

There may well be other factors that are relevant to explain the meaning of terms like 'available'. For instance, it might be that, in addition to the circumstances of the speaker, the conversational context matters. Perhaps there are contexts where we are only interested in evidence that is immediately ready to hand, yielding a parochial sort of chance, as well as contexts where we are interested in the sort of evidence that is available in the very limit of ideal inquiry, yielding a sort of idealised chance.

Variability in the meaning of 'chance'

Although there has not been much literature explicitly looking at the way chance claims can vary in meaning, depending on the circumstances of the speaker, there has been a great deal of philosophical and linguistic literature on a closely related topic: the analysis of *epistemic modal* terms, such as 'might' in the claim 'It might be raining in Morocco now'; or 'possible' in 'It is possible that it is raining throughout Western Victoria.' For further reading on the meaning of these terms, a good place to start is Hacking 1967, whose account of epistemic possibility has influenced my account of chance. Roughly, Hacking's idea of what is epistemically possible amounts to *that which is compatible with the available evidence.* So in working out the details of his proposal, he is engaging with the issue of what availability is. On this point, he writes:

A state of affairs is possible if it is not known not to obtain, and no practicable investigations would establish that it does not obtain.

Notice the force of 'practicable' here. We do not claim that no conceivable test would disprove our proposition. If determinism is true, and a Laplacian demon is conceivable, then some conceivable investigation would show before the draw exactly which ticket number would win [a lottery]. But no humanly practicable investigation could do that under the conditions of a fair draw.

(Hacking 1967: 149)

I believe my notion of available evidence fits quite closely with Hacking's idea of what practicable investigations would establish.

A further issue emerges, however, with epistemic modal terms, in that there are sometimes cases where apparently contradictory epistemic modal expressions can both be assertable by different speakers. You have not heard the weather forecast, and you say 'It might be fine tomorrow.' I have seen the latest satellite pictures, and say 'It cannot be fine tomorrow.' These assertions appear contradictory, but some theorists have thought that they can both be true – one common strategy has been to appeal to contextual variation to explain the compatibility of such sentences. I expect that much of this apparent context-sensitivity is relevant to the analysis of the term 'available' also, and by extension is of relevance to the characterisation of chance.

For a sophisticated contextualist proposal for modal terms that has influenced a good deal of philosophical discussion, see Kratzer (1991). Some of the important philosophical literature on epistemic modals includes: DeRose (1991) (this piece is not only the earliest I cite, but also the most accessible); Egan, Hawthorne, and Weatherson (2005); MacFarlane (2005); von Fintel and Gillies (2008); and Yalcin (2007).

There are some physical phenomena for which we are very confident that we know the chances with a high degree of precision. Here are two examples: first, as already noted, fair gambling devices are carefully manufactured so as to ensure that they generate particular outcomes with particular, commonly known, chances. So in the presence of a roulette wheel, we are – typically – capable of total confidence that the chance of getting any particular number, on a properly conducted spin, is $\frac{1}{37}$ (or, in the United States, $\frac{1}{38}$). Second, in certain physical experiments, we have extremely well-confirmed theories that allow us to make very precise probabilistic predictions about what will happen. We can seemingly know the chance that a sample of a radioactive isotope will decay to a certain mass within a given period, for instance. Or we can know the probability that one of a collection of subatomic particles, if measured for a property such as spin, will turn out to have spin up or down.

When thinking about these cases, one important thing we take ourselves to have learned is that there is a *limit* to what we can find out about the system, so as to enable us to predict what will happen. There are limits to what the available evidence can tell us about whether the next spin of the roulette wheel will land '23' or not. Indeed, it is part of the very purpose of designing a fair gambling device, such as a roulette wheel, that there be *no available evidence* that will make it possible, prior to the spin, to do better than to have a credence of $\frac{1}{37}$. Similarly, there are limits to what the available evidence can tell us about whether a particular radioactive isotope will decay. Despite our best efforts to try, we have not found any investigation or approach that is better than ascribing a certain probability to the event per unit time.[16]

There are other phenomena for which it seems foolish to imagine that we can know the chances with any degree of precision. Consider the proposition that the average global temperature will rise 2°C between the years 2000 and 2050. Some experts will have opinions to the effect that this is more likely than not. Others will think it 'very unlikely', and so on. But if someone confidently asserted that it had a chance of *precisely 0.43* of being true, we would rightly regard them as being somewhat unhinged.

My proposed account of chance may not explain everything that is suspicious about such a precise estimate of the chances, but it does go part of the way towards explaining it. In part, it is foolish for someone to make a claim like this because surely the speaker does not have *all the available evidence*. Moreover, if we were to obtain all the available evidence that bears on this matter, we would very likely have a different opinion as to the likelihood than we have now. By making a confident, and precise, estimate of the chances, this person seems to be suggesting that any further gathering of evidence that is currently available would not significantly alter this judgment. That seems obviously wrong.

The objectivity of what is available

One question, which arises from my analysis, is *whether the same evidence is available for everyone.* Marie Curie kept samples of radioactive elements in her pockets, oblivious to the dangers of ionising radiation. She died of aplastic anaemia. The following claim seems to be true:

Before Curie contracted aplastic anaemia, there was a high chance that prolonged exposure to ionising radiation (through, for instance, keeping radioactive samples in her pockets) would cause Marie Curie to contract some illness.

How do we know this? In large part, because we have learned a great deal from the evidence that has come to light *since, and because of* Curie's

---

16 Of course, it is no surprise to learn that many people have invested a lot of effort in trying to predict the outcomes of casino games. If one counts the cards that have been played so far in blackjack, it is possible to obtain a slight advantage over the house in the later hands. Hence casinos ban card-counting. They are trying to render that evidence unavailable. In the case of roulette, Thomas Bass has written an account of the somewhat mixed successes of a group who developed concealed computers which, when provided with some minimal input from an observer during the spin of a roulette wheel, would predict which octant the ball would land in with a better than $\frac{1}{8}$ chance (Bass 1985). Again, casinos do not allow such devices in the gambling halls, and so it remains the case for most of us that the best available advice is simply to apportion a credence of $\frac{1}{37}$ to each number on the wheel.

pioneering work. Arguably, that evidence was not available to Curie herself at the time.

If Curie herself was wondering about the possibility that radiation caused disease, she might entertain the thought:

There is a high chance that keeping this substance in my pocket will cause me to contract an illness.

This would be quite an *odd* thing to think, given what Curie knew at the time. But suppose, for the sake of argument, that she randomly hit upon this thought. We would like to say that, if Curie had entertained this thought, she would have stumbled upon a *truth*, even though she had no good reason to believe it at the time.

But is the following true?

The best identifiable advice function, given only evidence available now, recommends a high degree of belief that Curie will contract an illness if she keeps the substance in her pocket.

This too would be an odd thing to think. But is it *false*? If it is false, then it shows that Chance-4 is incorrect.

As I've said, I don't offer Chance-4 as an analysis. It might be that the notion of availability does not quite stretch to fit this scenario. My aim is merely to use a characterisation like Chance-4 to illustrate interesting features of chance. So, in that spirit, it is worth asking what sort of notion of 'availability' must be in play here, if I am at least approximately correct in thinking that available evidence is part of the meaning of a chance claim. Perhaps the sense of 'available' is not the straightforward meaning of that term in our ordinary language. But I claim that it is something importantly similar to that idea, and consequently that this is a helpful way of thinking about chance.

What we want is to allow that the evidence which allows us to learn scientific generalisations, such as the fact that ionising radiation damages the DNA in cells, and further, that damage to DNA in cells can cause cancer, is *always available*. Even though Curie could not in her lifetime complete all the experiments which led us to learn these generalisations, in some sense the evidence was 'available' because it was all around her: in nature itself.

But while, with regard to scientific generalisations, we are very inclusive in what sorts of evidence is available, there is also a danger in being too inclusive. If we allow particular *future* experimental outcomes to be included in the available evidence, then we will end up including in the available evidence propositions such as:

(4) Curie will contract aplastic anaemia as a result of exposure to ionising radiation.

And if that is included in the available evidence, then the best thing for Curie to believe about whether ionising radiation will make her sick is for her to be certain that it is true: to have a credence of *one*.

So to summarise: in general, chance is something like the advice given by the best advice function identifiable, using only available evidence. I'll abbreviate this, in future, to 'the best credence, given the available evidence'. Sometimes, in looser contexts, we might use chance talk with a very parochial notion of available evidence, and hence refer to the best credence, given evidence that is *immediately to hand*. This is the sort of context in which John can speak of the chance changing, merely because he has moved from his home to the doctor's office, thus making the test result more proximate. But in these looser contexts, chance loses its objectivity. For instance, when John is still at home, we would be required to think that the 'chance for John' is different from the 'chance for the doctor'. In other contexts, we use a different conception of chance, whereby the sort of evidence that is deemed available is something like 'practically possible to obtain'. Moreover, even this expansive understanding of availability needs to be stretched to the limit, to accommodate the idea that all laws of nature and similar factors are thought to be available, even to earlier scientists who we now realise were never practically capable of discovering those laws. This extremely expansive understanding of chance secures widespread sameness of chances for all occupants of large regions of space-time. Chances may be different at very different times and, correspondingly, chances may differ between two extremely widely separated locations in space, even at the same time. But this is not the sort of divergence that threatens objectivity.

Alternative conceptions of chance

Some philosophers – perhaps on the basis of the way in which chances can change upon receiving new evidence – will claim that I have erred already. What you have described is not properly *chance* (they will say). Rather, you have described something like 'objective epistemic probability'. That is not the same thing as chance. Chance is supposed to be irreducible, non-trivial probability in the world. It has *nothing to do* with what evidence we have. For example, chance and determinism are incompatible. If there are deterministic laws, then nothing is left to chance. But the way you have defined it, even in a deterministic world, if there is not enough evidence *available* to predict the future, you can say it is 'chancy'. That just shows you have not understood what we mean by 'chance'.

Jonathan Schaffer (2007) provides a good example of the sort of objector I have in mind. Schaffer not only disputes the compatibility of chance and determinism, he also claims that there is a conceptual connection between chance and the idea that the future is open and the past is closed. Hence he rejects the thought that past events can have chances less than one (p. 125).

This might appear to be a merely terminological dispute, but it goes beyond that. The dispute is also about which concept is more philosophically *fruitful*. I claim that mine is the more fruitful, in part because if you accept my characterisation of chance, it will do the work of *both* 'objective epistemic probability' and 'chance', as my objector understands these concepts. That is, my objector's idea of chance will turn out simply to be a special case of chance, as I have introduced it.

My objector understands 'availability' in the most expansive possible sense: the objector takes everything that we could *conceivably* know about the world at a particular time to be available. So it follows, for the objector, that chance and determinism are incompatible. Because if there are deterministic laws of nature, it is surely conceivable that we could have enough knowledge of the state of the world that we could predict the future with perfect certainty – even if *in practice* this is utterly beyond our powers. So the best thing to believe, given all available evidence, understood in this very expansive sense, for every possible future event, is to be certain that it happens or to be certain that it does not. No non-trivial chances will be required.

This special case of chance may be an important one. It is a concept that has come into widespread use among philosophers interested in chance in fundamental physics. But it is not the only concept of chance that we use. Moreover, by showing how the objector's understanding is just one instance of a more general concept that is in some sense epistemic, I can give a more unified account of chance and related phenomena. So my account is superior – or so I hope to show.

Before leaving this topic, it is worth briefly recapping the – admittedly slight – advantages of my conception established so far. Compared to conceptions that, like Lewis's, characterise chance as a trump, even over *complete* information about the past, my characterisation seems to better accommodate some of our ordinary thought and talk about chance. First, our ordinary chance ascriptions seem to be compatible with the possibility of deterministic laws. Second, if information about the future is inadmissible, then if I did receive a mysterious deliverance from a crystal ball, I could not revise my estimates of the *chances*, but instead if I change my degree of confidence I must say that I am doing so *contrary* to the chances. This too, seems odd.

A third advantage is that Chance-4 characterises the role of chance without reference to metaphysically substantive matters, such as the distinction between past and future, or the laws of nature. It seems plausible that I could communicate to someone about the chance of something, even where that person lacked the conceptual apparatus either to distinguish past from future or to distinguish laws from accidents. Of course, such a claim about the psychological means with which we carry out our chance talk is not the end of the story about what chance is, but it is a point in my favour if the notion of chance can be cashed out without appeal to additional substantive concepts such as these.

Having blown my own trumpet thus far, I now propose to stash it away for much of the rest of the book. Many of the interesting issues about chance are relatively indifferent to the subtleties of whether one accepts something like Lewis's characterisation or an alternative: the rough hallmarks identified earlier are enough to identify a target of common inquiry. So having pursued this dispute as far as I can to this point, I will, for the most part, proceed in later chapters without pressing the disagreement any further. One notable exception will occur in §25, where I take up some of these matters in more detail, primarily in response to Schaffer's articulation and defence of the Lewisian conception of chance.

5 What makes a fact of chance?

With this understanding of what it means to make a claim about chance, we might then ask: what makes it the case that the chance takes *this* value, rather than *that*? If the chance that P equals x, what makes it the case that the best degree of belief to have in P, given the available evidence, is x? This is the central question that we will be addressing in the remainder of the book.

There are already a number of philosophical accounts of what chance might be. We will look at some in coming chapters. But it is important to have in mind the central animating puzzle that motivates most of these accounts: that is, how do we account for the tension between the second and third hallmarks of chance? Chance dictates what we ought to believe, but it does not guarantee success. What then, is the relationship between a fact about chance and what *actually occurs*?

One natural way of thinking about chance is to think that chances are, at root, facts about mere *possibilities*. To say that something has a chance of happening is to say that it might happen. It is one way things could turn out. These facts about possibilities, moreover, cannot be reduced to facts about what actually happens. If we accept this non-reductionist approach to chance, how do we then answer the question about the relationship between objective probability and actual events? The obvious answer is something like this:

If an event has a high chance, then it is *very likely* to occur.

That is no doubt true. Perhaps we could use this claim as a starting point, proceed to generalise, and say some more about the relationship between probability and likely occurrence. But of course, this is a *circular* strategy. In trying to explain chance, we have used the term 'likely', and this is surely just another chance-like notion. So although what we said seems to be true, we have not managed to explicate the relationship between chance and what happens.

Some theorists, whom I call 'actualists', in virtue of being committed to various ideas about how we learn from experience, think that we must find a reductive relationship between chance and what happens if we are ever to make it explicable how we *learn* about chances. Since we learn by making observations of actual events, facts about chances must be reducible to facts about actual events.[17]

While it is true that there must be some relationship, it does not follow that the relationship needs to be the reductive type of relationship that the actualist offers. Note that when we try to learn about chances from observing what actually happens, it is part of the concept of chance that what we are doing is inelimably 'dicey'. That is: you can never be completely certain that, given an observed frequency of events, the objective probability of that event type is x, and no other value. If we observe 100,000,000 coin tosses,

---

17 Sometimes, some of the sorts of people I call actualists are known as 'Humeans' or 'frequentists'. But these terms are too narrow in scope. Actualism is just the idea that chances reduce to actual facts that are themselves non-chancy, in some sense. David Lewis, then, is an example of an actualist, even though he is famous for his belief in the reality of non-actual things (Lewis 1986a). Lewis is a non-actualist about the totality of what exists, but he is an actualist about chance, in my sense of the word.

and the number of heads and the number of tails is precisely equal, we still
cannot be sure that the coin is perfectly fair. At best, we can say: the chance is
very likely to lie within such and such a range. So when we estimate chance,
our estimates are themselves made in a probabilistic fashion.

In light of this, the request for a reductive account of the relationship
between chance and what occurs should not be treated as sacrosanct. We
need to recognise our own limitations when we try to discover facts about
chance. Given those limitations, a circular analysis of the relationship may
be all we could reasonably hope for. This would amount to some sort
of primitivist account of chance, whereby chance cannot be explained in
non-circular terms.

Primitivism about chance is not an untenable position. But it is sufficiently unattractive that it should be a final refuge, rather than an induce-
ment to early retirement. Primitivism of some variety may turn out to be
the best that we can do, but that is something to conclude only after we have
failed to do better.[18]

There are serious worries, then, about both an actualist account of chance
and a primitivist account of chance. One could simply declare it a philosophical tie: there are two distinct concepts in play, and there are no clear
grounds to prefer one over the other, as both suffer from problems.

But we do not need to use merely philosophical arguments to try to
understand chance. We can also look at the world, and in particular the
'deep structure' of the world, suggested by foundational theories of physics.
If a philosophical theory cannot explain the use of chance in a physical
theory, that will be good reason to reject the philosophy. Sometimes, we
will find that physics seems to be neutral: it does not favour one or the
other philosophical account. But sometimes physics can break us out of
philosophical complacency, and can suggest strikingly new ways to think
about what chance is. For this reason, this book will not be confined to
philosophical theories of chance, but will also examine – with a minimum
of technical detail – some of the ways chance arises in both classical and
non-classical physics.

The presentation here is aimed at an undergraduate student of philosophy: it presupposes no formal education in physics and involves minimal

18 I take myself to be only partly guilty of what Maudlin (2007b: 276) calls 'the usual
    philosophical move', which is to insist that the concept of primitive chance involves
    metaphysical questions or confusions which *demand* investigation. I do not claim that chance
    involves 'confusion'. I do think that positing primitive chance invites metaphysical questions,
    but I concede that there may be no illuminating answers to such questions!

mathematical detail. For many readers, more familiar with the physics, these chapters will involve rehearsing familiar territory, and can be skipped without significant loss. That said, my approach to these topics is one that aims to bring out the most philosophically significant features of the science, and thus may still be of interest, even to those with a significant background in physics.

The order of proceeding is as follows: Chapters 2–5 present the basic ideas needed to explain how chance might exist in classical physics – focusing in particular on statistical mechanics. In Chapters 6, 7, and 8, I adopt more traditionally philosophical approaches, and review possibilist, actualist, and anti-realist theories of chance. Chapters 9 and 10 turn back to physics, but with a focus upon quantum mechanics. In particular, in Chapter 10, I examine the case that has been made in favour of the many-worlds interpretation of quantum mechanics as a way to explain the existence of chance. Chapter 11 then revisits the relationship between time and chance, and attempts to explain the temporal asymmetry of chances. And finally, in Chapter 12, I return to the issue of realism about chance, and present what I take to be a decisive argument in favour of a limited anti-realism about chance.

2 | The classical picture: What is the world made of?

For a long time physicists regarded the world as conforming very closely to what I will call 'the classical picture'. The classical picture is most importantly based upon Newtonian mechanics, a beautiful and very powerful theory which is still used to make very accurate predictions about a wide variety of phenomena. But the classical picture is not restricted solely to Newtonian theories: by this term I mean a whole family of theories which were developed in the period before Einstein, and which were in broad agreement with Newton about fundamental matters. For example, what is known as classical electrodynamics is a theory that goes well beyond Newtonian mechanics in the sort of phenomena it describes, but it is still recognisably part of the classical picture.

In the following sections, I will present a heavily simplified version of the classical picture. Many of the ideas will be familiar to many readers, and it might seem unnecessary to rehearse them. It is often unappreciated, however, that these features interact to generate an *overall conception* of the world. To make the classical picture vivid, I will ask you to imagine a fictitious being, much like a deity, attempting to record in perfect detail exactly how the world is at a particular time. Suppose this deity wanted the perfect memento of the world at a given time, and had unlimited resources and ability to record all the information about the world at that time. I don't mean to suggest that such a being exists. By focusing on what such a creature would *need to do* in order to record all the details of the world, however, we can give due emphasis to the most interesting features of the world as a whole.

The purpose, then, is to draw attention to some very general facts about the world, and to put them into a 'God's-eye perspective'. You might be anxious that the view from this perspective is very different once we adopt more contemporary physical theories as the basis for our inquiry. That anxiety is quite legitimate, and will be addressed in later chapters, when I will revisit the claims made here and revise them in light of contemporary physics. For now, however, you should bear in mind two things. First, many of the things that are true and interesting (especially from a philosophical perspective) about classical physics are also true in non-classical physics. Occasionally some details need tweaking, but having made those adjustments, a recognisably similar idea is true. Second, much of our common-sense thought is already quite challenged in classical physics. There are many more challenges to come with non-classical physics, and it will be easier going if we save those for later.

6 Matter is made of particles

On the classical picture, the world is made up of particles in space. That might seem obvious, for many of us have been brought up to think of the world in precisely these terms. But if you pause to think about it, it is quite contrary to our everyday experience to think of the world as made up of particles. Rather, many of the substances you and I deal with appear to be made of matter which is divisible into smaller bits of the same substance, and which has no obvious 'granularity'. If I kept dividing it – you might think – I could just keep going forever, making ever smaller portions. Indeed, every time I do divide it, what I get is precisely the same stuff, just differently arranged in space.[1]

Take a bowl of water. I can separate it into two smaller bowls. Then I could divide one of the smaller bowls into two cups. A cup could be split into three eggcups. An eggcup could be split into ten teaspoons. A teaspoon could be split into four quarter-teaspoons. Each quarter-teaspoon could be split into a number of droplets. Each droplet could be split into even smaller droplets. And so on.

Eventually, the portions of water will be so small that I am unable to divide them further. But that will not be – it appears – because I have got down to fundamental indivisible particles of water. Rather, it will just be because the tools I have are too big and clumsy for the tiny speck of water with which I am dealing. Moreover, the tiny speck I have will still behave in very similar ways to the original bowl of water. It may evaporate – so it may no longer be in liquid phase. But it will remain a tasteless, colourless substance that does not mix well with oil, etc. So I might well think that, in principle, it is possible to keep dividing the water indefinitely.[2]

---

1 There are some branches of classical physics, such as electrodynamics, which embrace non-particulate phenomena. Ignoring such phenomena is one of the ways I am simplifying the classical picture.

2 Contemporary metaphysicians have taken to calling substances that can be infinitely divided: 'gunk'. The idea is a very old one, however, going back at least as far as the Stoic philosophers.

On the classical picture of the world, however, we take the appearance of infinite divisibility to be an illusion. The difficulty I have in dividing the water forever is not merely practical. There is *in principle* an obstacle to my ever dividing the water beyond a certain point: namely, that the water is made of molecules, and the parts of the molecules are not themselves 'watery'. That is, if I were to split a water molecule up into its component parts, I would get hydrogen and oxygen atoms. And once split apart, these things are simply not water. Neither hydrogen atoms nor oxygen atoms – taken in isolation – are tasteless, colourless liquids with the properties of water. Indeed, even a single molecule of H_2O will lack many of the properties we ordinarily associate with water, since those properties arise from the interactions between molecules of H_2O.

One way to put this thought is that water has a 'fundamental' level of existence.[3] The molecules are the most basic unit you can get that is still water. Decomposing the molecules is possible. But once decomposed, you don't have water any more. So for the purposes of water's existence, the molecules are fundamental.

And what goes for water goes for everything. Substances which seem to be simple and infinitely divisible are in fact, at their fundamental level, arrangements of particles. Moreover, some of the constituent particles are themselves entirely basic. They are true atoms in the sense of 'things without parts'. They cannot be broken down further: they are simply the most basic form of matter.

Current physical theory suggests that the fundamental particles are things like electrons and quarks. But the classical picture does not require that we believe in these particular particles. The key point is that the classical picture involves the idea that *all matter is fundamentally particulate*.

7 Particles have properties

Not all particles are the same. Some are heavy – or, to be precise, more massive – and some are less so. Some are electrically charged, and others are not.

The term appears to have been introduced by Lewis (1991). Nolan (2006) argues that Chrysippus advocated a metaphysics of gunk.

3 Jonathan Schaffer (2003a) has argued provocatively that there is no good evidence to believe that there is a fundamental level of reality. Nonetheless, it seems that we can make sense of the idea that there is a fundamental level at which any given kind exists. It is to this idea of relative fundamentality that I am appealing here.

These properties are important because they affect how the particles will move over time. For instance, two massive particles will, other things being equal, move towards each other. Two particles that have similar electrical charge will, other things being equal, move away from each other.

One important thing about these properties which makes them relatively simple to keep track of is that they do not change with time. So we can call these the *static* properties of the particles. Any other properties of the particles will be *dynamic* properties: those which do change with time.

What are the dynamic properties? Simply the *positions* of the particles – nothing more. (It is hard to imagine a simpler theory!)

How, then, would the memento-making deity go about recording this information? A very natural way would be to write a list. Such a list would contain an entry for every single particle. It would be convenient, then, for the deity to label the particles in some way, so that each item on the list is associated with a label. Given that there may be very many particles, the simplest way to label them is probably just to number them. So we have a numbered list, where each numbered item corresponds to a distinct particle.

To complete each entry, then, the deity will need to record the properties of each particle. Next to each particle we can write down its mass, its charge, and any other static properties that we might need to record. We can also record the dynamic properties of the particles: that is, the positions of the particles in space.

Assuming a very simple-minded way of thinking for a moment, this could seem problematic. How can you describe a position in space? In particular, because space is continuous, there are infinitely many positions in space. So how do we write down descriptions that uniquely pick out each and every location?

The solution is to use a coordinate system. We typically use what are known as Cartesian coordinates, but there are other possible coordinate systems which we could use. In a three-dimensional space, we can systematically label each and every point in space using three numbers. Each number represents the distance, in the relevant dimension, from a specified origin. The great advantage of a coordinate system is that you can use it to give a unique description of infinitely many points and these descriptions can be systematically related to each other.

In Table 2.1, I have given a basic example of what such a list might look like, for a two-dimensional space. Note how useful numbers are in constructing this list. Throughout, we use numbers. Numbers describe the positions of the particles, and they also describe the static properties of the particles.

Table 2.1 A rough sketch of what our deity's list might look like, if particles had only mass and charge as static properties, and if space were two-dimensional

#	Mass	Charge	Position (x)	Position (y)
1.	1 gram	+1	13	−4
2.	10 grams	−2	3	14
3.	3 grams	0	5	0
⋮	⋮	⋮	⋮	⋮

In the mass column, I included the units of measurement to accompany the numbers. That is, I specified that the mass was a certain number of *grams*. But of course, for all the entries in the table – for charge, for position, and for any other properties we might need – we would need to be using units also. I have simply omitted them for convenience.

While numbers are very convenient representational devices for these purposes, one must be careful not to attribute more meaning to these labels than is intended. In a number of ways, the use of a number is in some sense *conventional*. That is, the meaning of the number is governed by rules that we have established. If we do not pay attention to those rules, we can misunderstand the meaning of the numbers.

For instance, it is appropriate, given the figures in Table 2.1, to say that particle 2 is more massive than particle 1. To make such a claim is to use the convention regarding representation of mass correctly. It would not be appropriate, however, to say that particle 2 'has a larger position in the *x*-dimension than particle one'. This is a mistake because the figure in the *x* column is not a 'size' of a particle. It is the magnitude of the displacement between the particle and the origin in the *x*-dimension.

Here is another example. If a particle had an entry of zero in the mass column, it would be appropriate to say of it that it has 'no mass'. But if a particle had an entry of zero in one of the position columns, it would be absurd to say it has 'no position'. It evidently does have a position, but that position coincides with the position of the origin in the relevant dimension.

It is safe to say that the attempt to label points in space using a coordinate system is typically subject to more errors of interpretation – or at least subject to more obvious errors of interpretation – than our attempt to describe static properties like mass and charge. We will have more to say about this in later sections.

8 The laws are deterministic

In the classical picture, the laws are deterministic. Put roughly, this means that the state of the world at one time determines the state of the world at later times. The laws entail that there is only *one possible way* that the world could evolve in future, given its state at a particular time.[4]

Here is one way you might interpret that claim, in the context of our story about the memento-making deity. The deity has a perfect record of the instantaneous state of the world at a particular time. The deity, being very intelligent, also has the ability to perform calculations, using the laws, to predict how this state will evolve in time. Because the laws are deterministic, the deity can therefore deduce, simply from the laws and from the recording, every single future state of the universe.

This is actually not correct. The memento and the laws together do not quite suffice to determine the future evolution of the world. The thing that is missing is that the memento does not include the *velocity* of the particles. Velocity, of course, is a measure of how fast a particle is changing its position over time, in a particular direction. Recall that, in the list that we suggested the deity would make, we recorded only the static properties like mass and charge, and also the positions of the particles. We did not explicitly record the rate at which the particles were changing position. And without this information, the laws will not be able to tell you how the world will evolve in the future.

This might seem odd: surely the laws determine, on the basis of things like charge and mass, how fast things will move?

What the Newtonian laws determine directly is *acceleration*. That is, they determine the rate at which velocity is changing. But to take that information and turn it into information about where a particle will go, you need to know how fast it was moving in the first place.

This is worth rehearsing in a little more detail. The Newtonian laws will tell you, for instance, that two objects that have a particular mass, and which are a particular distance apart, will be subject to an attractive *force* of a certain quantity. The law in question looks like this:

$$\text{Attractive force} = \text{Gravitational constant} \times \frac{\text{Mass}_1 \times \text{Mass}_2}{\text{Separation}^2}$$

4 Some potential counterexamples to the claim that classical physics is deterministic have been pointed out by John Earman (1986: chap. 3) and John Norton (2008). These complications can be safely ignored for our purposes.

Table 2.2 The deity's augmented list, including instantaneous velocities for all particles

#	Mass	Charge	Position (*x*)	Position (*y*)	Velocity (*x*)	Velocity (*y*)
1.	1 gram	+1	13	−4	2	−4
2.	10 grams	−2	3	14	0	18
3.	3 grams	0	5	0	1	1
⋮	⋮	⋮	⋮	⋮	⋮	⋮

That is, the attractive force between any two massive objects is equal to the product of those masses, divided by the square of the spatial separation between them, and multiplied by the gravitational constant.

Forces are related to movement by Newton's second law, which states that:

$$\text{Force} = \text{Mass} \times \text{Acceleration}$$

So if we know that an object is subject to a force of 6 units, and it has a mass of 2 units, then we can solve the equation to deduce that it is being accelerated by 3 units of acceleration.

But for all that, the object could be stationary; *or* it could be moving at an enormous speed in one direction, and the acceleration would be speeding it up; *or* it could be moving at an enormous speed in the opposite direction, in which case this acceleration would be slowing it down; etc. Until we know the velocity of a particle, we cannot use the Newtonian laws to work out how it will move, even if we know about all the forces being exerted on the particle.

So if our deity wished to record a slightly more detailed memento, one which includes not only a perfect description of the world at a particular instant, but also allows the deity to deduce all the future states of the world, then we need simply to add some extra columns to our list. For each dimension of space, we need a velocity column.[5] So if we stick with our simple two-dimensional space from the earlier example, the augmented memento might look like Table 2.2.

---

5 There has been debate about whether or not instantaneous velocity is itself an intrinsic property of an object at a given instant, or if it is fundamentally a property that obtains in virtue of relations between different times. I am supposing it is something like an intrinsic property, but for present purposes nothing important turns on this. See Arntzenius (2000); Bigelow and Pargetter (1989); Tooley (1988).

9 And that's all there is

Particles changing position over time.

That, in a nutshell, is the entire history of the world according to the classical picture.

This is a spectacularly ambitious claim. To get a sense of the ambition involved here, consider a process that we have good reason to believe is happening all over the world many billions of times a year: the metamorphosis of a butterfly. A wormlike thing, a caterpillar, builds a cocoon, and while residing inside it, undergoes a transformation into a very different sort of creature, quite unwormlike, with wings and often dazzling colours.

If you wanted to describe this process in ordinary terms, you would presumably be interested in the sorts of change required to transform the tissues that form the body of a caterpillar into the tissues that form the distinctive body of a butterfly. How do you go from green caterpillar-body to bright turquoise wing-patches, for instance? It is possible, of course, that the turquoise wing-patches were always hidden inside the body of the caterpillar, and metamorphosis simply involves an elaborate rearrangement of these parts. But we now know that this is a rather fanciful idea, and that some sort of synthesis occurs to grow the wings, rather than simply reassemble them.

But in another sense, this fanciful thought is perfectly correct. If the classical picture is right, the entire process of metamorphosis is simply a rearrangement of fundamental particles in space. Those particles do not themselves undergo *any* intrinsic change. If they were all heavy, positively charged particles to start with, they are all heavy, positively charged particles at the end of the process. The *only* thing that changes is their arrangement in space.

Of course, the things that are arranged are much smaller than gross body-parts of a butterfly. Indeed, it is very hard for us to conceptualise just how many particles are involved in constituting a butterfly: something in the vicinity of 10^{23}. (That is, a one followed by twenty-three zeros.) So it should not be so surprising, perhaps, that mere spatial rearrangement of otherwise unchanging particles could bring about such a radical change, given we are so poorly acquainted with the particles themselves, and relatively incapable of conceiving just how complex their arrangement really is.

What is more, the spectacular ambition of physics to account for *every* phenomenon of the world has not changed since the time of the classical picture. The non-classical picture is a bit less easy to describe, but it is still stunningly spare and unfamiliar; and it remains a shock to suppose that physics, in some sense, is all there is.

Of course, you might accept all of what we have said so far regarding the nature of particles, of matter, and of how matter changes with time, but you might resist this final thought that 'that's all!' Many people are inclined to think that mental phenomena, in particular, are flatly inexplicable in terms of physical processes like those of the classical picture. Consequently, they are often tempted to think that there must be something else. This motivates many to adopt a sort of dualist account of the world: an account which maintains that there are two fundamentally distinct types of thing in the world – mental stuff and physical stuff.

I won't discuss the plausibility of dualism here, but I will say something about the ambition of physics. Physicists have good reason to believe that everything that happens in the world, be it the metamorphosis of caterpillars, the performance of a play, the decline of an economy, the evolution of a new species, and even the contemplation of an interesting idea, is at least compatible with the physical laws. So there is no reason, it would appear, to think of non-physical phenomena as somehow 'interfering' with the operation of physical laws. There is no evidence to suggest that there are any non-physical phenomena – be they mental forces, economic activities, or supernatural influences – that render the laws of physics false or 'non-operative' in some way. The most plausible way to resist the idea that physics is complete, then, is to say that physical laws *leave various possibilities open*, and that the other phenomena have a role in influencing which of the physically possible outcomes eventuates.

Part of the problem with this way of thinking – at least on the classical picture – is that there does not seem to be much room for physics to leave things 'unsettled'. Recall that the laws are completely deterministic. So at least as far as movements of physical particles go, there simply are no alternative outcomes that were physically possible, but which did not eventuate because of the influence of some additional factor. One could try to hold on to the idea that there are non-physical factors involved by supposing that they do not operate on the physical particles of the classical picture, but rather on some other, non-physical stuff. The problem for this way of understanding the world is that it becomes deeply mysterious how the physical stuff could interact with the non-physical stuff. Again, the laws governing the physical particles are wholly deterministic, so how could an interaction with some other stuff cause it to do anything that it wasn't already determined to do?

This is just a brief sketch of a long tradition of debate relating to a whole host of special sciences which seem to have some autonomous status – in that they seem to have laws that are very different in form from physical laws, and they seem to make little or no reference to the fundamental categories of physics. On the other hand, however, these special sciences are hard to understand as truly independent, given it is implausible to think that ecosystems, economies, and chemical reactions are not ultimately constituted by purely physical processes.

I cannot hope to settle these debates here. For now, I invite the reader to entertain the classical picture in its full ambition. Suppose the history of the world really is nothing more than particles changing their positions over time. Even if you think that such a world would lack minds, or economies, or ecosystems, and therefore that we do not live in a world like this, you can still ask what such a world would be like. And, most importantly for present purposes, we can ask what *chances* would be like in such a world.

10 Do the laws heed the direction of time?

When I spoke about the determinism of the classical laws, I stressed the way the state of the world at a particular time, plus information about the velocity of the particles in that state, determines the *future* states of the world.

This is entirely true, but it is misleading, because it suggests that the determining power of the laws operates in one direction only: the 'forwards' direction of time. What is actually the case is that the laws determine the totality of what occurs, in *both* directions of time. Given a memento of a single moment, augmented with velocity, the laws determine the complete state of the world at all past times and for all future times. So this is one way in which the laws do not seem to heed the direction of time. They do not determine the state of the world in one temporal direction, while neglecting the other. Rather, they appear to determine the state of the world in both directions.

There is, however, another way in which the laws do not seem to heed time in the way we might have expected them to. Consider the augmented recording in Table 2.2. Now, we might wonder about what the world would be like, had things been a little different from the world as described in that record. We might wonder how things would go if particle number 2 had been slightly closer to particle 3. Or we might wonder how things would go if particle 1 was moving faster in the *x*-dimension. These alternative possibilities that we might wonder about correspond to different ways of filling in the table. Indeed, each way of filling in the table seems to represent a different way the world might have been.

Here is one especially interesting way the world might have been: the velocities of all the particles, in all dimensions, might have been precisely the opposite of what they actually are. So for a particle which currently has a positive velocity of +3 in a given dimension, it could have a velocity of −3 in that dimension. And so on for all particles in all dimensions.

To generate a full description of this possibility from Table 2.2 would be a straightforward matter; you would simply need to multiply all the figures in the velocity columns by negative one.[6]

Suppose we did this, and came up with a new description of the world in the same tabular format as the deity's memento. Call this recording *Opposite*. The world could have been just as we have described in Opposite. If that had been so, how would the future look, compared to the actual world? It would look exactly like the recent history of the actual world, *in reverse*. Let's consider this in some detail.[7]

First, let's make some suppositions about the actual world. In the actual world, one hour ago, I walked into my office and got out my coffee mug. I then went to get some coffee from the common room. I then started to drink the coffee, while typing this section of the chapter. By the time the deity records the complete state of the world, the coffee is three-quarters gone, and what coffee remains is a good deal cooler than it was when I first made it. There are also many more words in this document than when I started typing less than an hour ago.

In the world described by Opposite, the future is utterly bizarre and alien compared to the sorts of processes with which we are familiar. A person whose body contains particles arranged very similarly to mine sits at a computer keyboard. Periodically, he undergoes a process a bit like regurgitation, bringing up mouthfuls of coffee and depositing them in a cup. Moreover, the coffee in the cup keeps getting hotter. It absorbs heat from air molecules that bump into it, and the overall air in the room gets very slightly cooler, while the coffee actually heats up.

The document on the computer, meanwhile, is rapidly unwritten. The characters continue to disappear from the screen, with occasional re-appearances, coinciding with when the typist hits Delete. With each

6 If we enriched our understanding of the physics a little, to include electromagnetic charge, then in addition to reversing the velocities we would also need to reverse the direction of the magnetic field. See Albert (2000: 20–1) for discussion.

7 See Albert (2000: chap. 1) for a more thorough discussion of the time-reversibility of physical theories.

movement of his hands, this person seems to be engaged in an activity like reverse-typing. As each letter disappears from the screen, the typist very shortly thereafter hits precisely that letter on the keyboard.

In short, this world would appear like a movie of the actual world, played in reverse. And when you think about that, you soon realise that the world is full of many processes which normally occur only in one temporal direction: from past to future. To suppose that those sorts of processes could occur in the opposite direction seems utterly fantastic.

To help you to comprehend these strange events, I have helped myself to the idea that it would 'look' a certain way. But even to do that is to make this world more familiar than it really is. If you walked into my office right now, watching me type, you would see me because light coming from the sun and from the fluorescent lamp above me would reflect off my body and clothes, would be focused by your eyes onto your retina, and would stimulate your brain appropriately. But in the world described by Opposite, light would typically go *from* a person's retina, out to the object they are 'looking' at, and then reflect off that object back to the light 'source'. In other words, nothing remotely like normal visual processes would occur in this world. So to suppose that it would 'look' like anything at all familiar is probably incorrect.[8]

However, the sober conclusion we must draw from the laws is that this possibility is just as viable as the actual world in which we take ourselves to live. The laws of Newtonian mechanics do not distinguish between the possibility of a given process occurring in one direction in time, or a very similar process – different only in the velocities of the particles – occurring in the opposite direction in time.

This feature of the laws is known technically as 'invariance under time-reversal'. More loosely, it is also referred to as the time-symmetry of the laws. What it means is that for any possible continuous sequence of physical states, S_1 to S_n, there is an operation to be performed on those states – such as multiplying the velocities of the particles by negative one – which will generate another possible sequence of physical states, S_1^* to S_n^*. And this new sequence of states will have the same number of particles in the same positions as in the previous sequence, but the order of them will be *time-reversed*. That is, S_n^* will be the earliest state in the sequence, and S_1^* will be the last.

8 This argument is due to Maudlin (2007a: 121–4).

The same goes for other classical theories. The operation to be performed varies, but the essential idea is the same. As far as the positions of particles is concerned, *precisely the same process can happen in either temporal direction.*

Needless to say, this feature of classical mechanics is profoundly at odds with our everyday experience. To give just a very brief list of the sorts of processes that seem to occur in only one direction: the hatching of a chick from an egg; the falling of ripe fruit from a tree; the eruption of a volcano; the cooling of a hot meal left on a table; the slow leak of air from a tyre; the radiation of light from the sun; and so on, and so on. It is not just that these processes would be very surprising if they happened in reverse. Rather, the naive view is that the reversed versions of these processes are physically impossible.[9]

So there are two important ways in which the laws seem to ignore the direction of time. The first is that the laws determine what will occur both for the future and the past, given a complete description of the world at a particular moment, including the velocities of the particles. We typically think of the laws as producing future states from past ones, not as determining past and future alike.

The second way is in the time-symmetry of the laws. All processes, according to the classical picture, could occur in the opposite direction in time. This second feature, in particular, seems flatly contradicted by thousands of years of human observations to the effect that some processes happen in one temporal direction only.

---

9 The attempt to reconcile invariance under time-reversal with our experience of the world as full of processes that happen in only one temporal direction is quite a sizable topic that bears on the nature of chance. We will revisit this matter in Chapter 11.

3 | Ways the world might be

Obviously, our knowledge of the world is quite unlike our hypothetical deity's. We don't know exactly how many particles there are. We don't know their exact properties. We don't know their exact positions.

One important way to distinguish our state of knowledge from the deity's is that for us, there are *lots* of ways the world might be. If we were shown the deity's memento, even if we had time to examine it, we still could not say with confidence that it is an accurate representation of the world.

This suggests a link between knowledge and possibility. Our state of knowledge is one that is compatible with a number of different ways the world might be. And this is a key difference between our state of knowledge and the state of a being who knows everything. Such a being is in a state that is compatible with only one way the world is. In this chapter, I will explore this link further, and consider how we can represent certain sorts of knowledge in terms of ways the world might be.[1]

11 A multitude of lists

The deity's memento is a perfect record of the world at a particular time. It consists simply of a table that lists every particle and the physically important properties of each particle: mass, charge, position, velocity, and any others that might feature in the laws.

We have already, in the previous chapter, invoked a very natural idea about how to understand alternative possibilities in terms of this table. The table as it has been filled out represents the way the world is. *By filling out the table differently, we represent an alternative way the world might have been.* In the previous chapter, this idea was employed when I suggested you could

---

1 Note that when I talk about ways the world might be, I am not merely talking about ways the world might *come to be*: I mean to include ways the world might have been in the past, might be now, or might come to be in the future.

Table 3.1 Two tables, each representing the way a world might be

#	Mass	Charge	Position (x)	Position (y)	Velocity (x)	Velocity (y)
			(a) First table: a three-particle world			
1	2.09	−2	3.5	2.6	193	−234
2	12.2	+2	123	1.4	0	31
3	456	−1	−67	−902	0	0
			(b) Second table: a one particle world			
1	2.09	−2	3.5	2.6	193	−234

multiply all the numbers in the velocity columns by negative one, and thus generate the velocity-reversed image of the actual world. Such a world was, we assumed, possible. And indeed, the classical account strongly suggests it is.

To try to get a grasp of *all* the possibilities then – a grasp of all the ways the world might be – we need to generalise the idea of changing the numbers in the table. Here is a first, very rough, attempt at making that idea explicit:

P₁ A possible way the world might be is simply a list, like Table 2.2, with some number N of rows, and a particular number of columns, corresponding to the number of properties each particle has. Each row of the table is numbered using the numbers 1 to N, and each field in the table is populated with a number.

Table 3.1 shows two very simple tables, each of which, according to the above principle, is a way that the world might be.

Notice that one very obvious way in which these worlds differ is that they contain different numbers of particles. The first contains three particles, while the second world contains a solitary particle. This seems right: the world could have had more or less particles than it actually does. For some purposes, perhaps, we might only be interested in other ways the world could be, *supposing the number of particles to be the same as the actual world*. But that would simply require us to restrict the tables of interest to those with the same number of rows as the deity's memento. There seems no reason to build this restriction into our very definition of a way the world might be, however.

One concern we might have with the proposal P_1 is that it could generate too many possibilities. For instance, consider the alternative table you could get from Table 3.1(a) by switching the labels 1 and 3 in the '#' column. That is, particle #1 would now be a very heavy, stationary particle, and particle #3 would be a light, fast-moving particle, rather than the other way around. Is this really a different way the world might be? When we consider possible ways the world could be, do we really think something could constitute a different way the world might be, not in virtue of what properties the particles instantiate, but in virtue of what particular particles instantiate the various properties? (The philosophers' term for such differences is a 'haecceitistic' difference between possibilities. This term comes from the Latin for 'thisness'. The difference is not in any of the qualities present in the world, but in the particular thisness of the particles that exemplify the qualities. If you don't believe in haecceities, then you are not likely to believe in merely haecceitistic differences.[2])

Philosophers are frequently sceptical of such differences, on the grounds that there would presumably be no *observable* difference between Table 3.1(a) and an alternative in which the labels 1 and 3 have been switched around. That certainly seems a sensible attitude to take, but I would urge a more hesitant approach for the present. We have not yet discussed the way in which alternative possibilities are used in our scientific activities. Once we have a better grasp of the *role* of alternative possibilities, we will be in a better position to decide whether or not to acknowledge such obscure differences between possibilities.

So, for now, I shall leave it unresolved whether or not the labels on the rows matter. It could be that we should simply understand an alternative possibility as an unordered table – as nothing more than a collection of rows, and thereby deny that there is any difference between the collection $\{A, B, C\}$ and the collection $\{C, B, A\}$ – where A, B, C are rows of a table. Alternatively, it could be that the identity of the particles is partly constitutive of the identity of an alternative possibility, and therefore that the order of the rows tracks a genuine feature of each possibility.

---

2 The classic thought experiment which pertains to the existence of haecceities involves two qualitatively identical black spheres (Black 1954). If nothing else exists in the world, is it necessary to suppose that each sphere has its own 'thisness' to make it the case that the world contains two spheres rather than just one? An important discussion of haecceitism, in the context of distinguishing possibilities, occurs in David Lewis (1986a: §4.4).

12 Possibilities that differ spatially

There are still more reasons to be worried that the proposal P₁ will generate more lists than there are genuine possibilities. Consider the following. Start with Table 3.1(a). Now add two to all the figures in the column for position in the *x*-dimension. You will get a different table. Is this table, however, a genuine alternative possibility?

Supposing it is an alternative possibility, then this world is just like the first, only it is 'moved over' two units in space. Is that really a *different world*? Once again, you seemingly could not conduct any *experiments* to determine which of the two tables better describes the world you are living in. Therefore, you might think, there is no real difference here.

Recall, the figures in the position columns represent *displacement* from an arbitrarily chosen origin. There is no glowing point somewhere in the middle of the universe – or anywhere else for that matter – that *must* be used as the origin. Rather, we need to pick somewhere – anywhere – as an origin, in order to give us the means to label all the points in space in a systematic fashion. As long as we consistently use the same point, we have no troubles. The numbers seem, then, to be tracking position relative to an arbitrary choice of origin. There seems to be no good basis in our coordinate system for thinking that the two lists described above are really different worlds.

(Contrast what the world would be like where you added two to all the figures in the *mass* column of Table 3.1(a). This would be a manifestly different world, with significantly heavier particles. There would be utterly different forces between the particles, and consequently a very different past and future.)

Even though the coordinate system itself does not *require* that there is any real difference between these lists, nor does the system rule it out. It is conceivable that there really is such a thing as a purely spatial difference between two worlds, even though we have no way to ensure that the way we use coordinates in the actual world 'lines up' with the way they are used in alternative possibilities. So again, as philosophers we should be cautious that we are not too dismissive. It might turn out that worlds that differ in such ways can play a useful role in scientific inquiry. If that is the case, then we should not insist, on the basis of philosophical reasoning, that such differences are spurious.

Issues such as the one we have been considering are usually considered as part of a bigger debate about the nature of space itself. On one account, known as relationalism, space itself is not a substantial thing, capable of existing independently of *things*. Rather, space is what you get when there are things that stand in spatial relations to one another. Therefore, this view entails that there cannot be differences between worlds in virtue of transformations like adding two to all the *x*-coordinates. The second list would involve the same spatial relations between particles, and that is all there is to space. So there is no difference to be found.

The opposing view is known as absolutism about space.[3] Absolutists maintain that space is a substantial thing that exists independently of objects standing in spatial relations. It is in virtue of things occupying certain locations in space that they get to stand in spatial relations to one another. Therefore, by adding two to the spatial coordinates of all the particles in a world, you would be representing a genuine change in location of all the particles. The space would be the same, but the particles would be located differently in that space. Therefore the two lists would indeed be distinct possibilities.

As it stands, the proposal P_1 seems to assume something like absolutism. To re-draft our proposed account of alternative possibilities, so as to have it favour relationalism, we would need to come up with a general principle that identifies tables that differ only in terms of simple, global transformations upon the spatial coordinates, in one or more dimensions. For instance, we would want to cover transformations which represent *rotations* of all the particle positions; *reflections* of all the positions; and simple *displacements* of all the positions. Let's just call these transformations the 'global spatial transforms'. We would then want to take each list and put it into a group with other lists that differ from it only by one or more global spatial transforms. What we will get, then, is all the original lists, clustered into what mathematicians call equivalence classes on the relation of a global spatial transform. Each equivalence class would contain all the different ways we could assign spatial coordinates to what is in fact just *one* possible way the world might be. We could then say that:

P_2 A possible way the world might be is an equivalence class formed from a list, like Table 2.2, under the equivalence relation of global spatial transformation.

What we have done by using equivalence classes is simply to *ignore* some of the structure of our tables. If we are relationalists about space, we must be of the view that the original tables contain more structure than is required to represent a way the world might be. In order to get at the relevant structure,

---

3 The most famous debate over relationalism and absolutism was that between Gottfried Wilhelm von Leibniz and Samuel Clarke. Their correspondence is published in a number of editions.

we simply group together multiple tables which share the relevant structure, and differ only in ways that are irrelevant.

I am not going to attempt to settle the question of absolute versus relational space. Rather, the point of this exercise has been to show that, if we are going to come up with a theory of alternative possibilities, we will need to be alive to the danger that our representation will contain unwanted structure. However, there is a very useful technique available to eliminate such unwanted structure, and I have given a very brief demonstration of how that technique can be used.[4]

13 Pushing the limits of possibility

The proposals in P_1 and P_2 are both relatively neat, and seem to cover plenty of possibilities. They allow that there could have been any number of particles, that they could have had any combination of properties, and – subject to caveats about the debate over absolute and relational space – allow the particles to occupy an enormous range of possible spatial positions.

It is worth stressing that the number of possibilities one thereby gets is *infinite*. Indeed, it is very easy to get infinitely many possibilities. Consider a very simple world, with just one particle, as in Table 3.1(b). Look simply at that particle's mass. If its mass can be any positive number, then already, just focusing our attention on one-particle possibilities, we have infinitely many ways the world might be. For the particle could be 1 gram, 2 grams, 3 grams, 3.7 grams, 3.7000001 grams, or 3×10^{100} grams, or ... There is literally no end to the possible masses the particle might have.

The same trick can be applied to the particle's charge. So with respect to charge, there are infinitely many possibilities for the particle. Then, of course, the possibilities for mass and the possibilities for charge can interact, giving what we might naively call 'infinity times infinity' many possibilities. In many distinct dimensions then, for each particle there are infinitely many possibilities.

Surely this is enough, you might think! How could we want more possibilities than this? However, recall why we began this discussion of possible ways the world might be. It was because we thought there was a link between possible ways the world might be and our knowledge of the world. Given that, there is in fact reason to be concerned that proposals like P_1 and P_2 do

---

4 Earlier, when discussing the possibility of haecceitism, I demonstrated an even simpler method of eliminating structure: simply eliminating some of the information from the table.

not in fact capture all the possibilities that we need. The reason is that the tables we have been using to denote possibilities all *agree* on some fundamental issues. They agree (i) that matter is all particulate, and (ii) that the only properties of things are mass, charge, position, and the like.

For the purposes of a physicist, this agreement is rarely going to be troubling. Provided you are happy to assume that the classical picture is correct, then you will indeed only be interested in possibilities that satisfy these constraints. But sometimes we want to entertain more outlandish possibilities. Isn't it possible that the moon is made of some substance that is not infinitely divisible, and therefore not particulate? Could there be real ghosts, made of ectoplasm, which cannot be characterised in terms of the fundamental properties like mass and charge? Could we be wrong in thinking that the number of particles stays the same over time – maybe they just pop out of existence occasionally?

Speculations like these might seem foolish and unimportant, but even practising scientists will sometimes want to consider the possibility that their theories are incorrect. Consider three doubts a scientist might have about her current theory:

1. Matter might not be all particulate.
2. Gravity might not obey an inverse square law.
3. It might be that our current list of physical properties is inadequate. It might need to include the strange sorts of properties described by the wave function of quantum mechanics, for example.

These doubts seem to raise alternative possibilities, in some sense. And part of the job of a scientist is to work out which possibility we should believe in as real. The problem for proposals like P_1 and P_2 is that not all of these possibilities can be represented simply by changing the numbers in a list like the ones we have considered so far.

Consider the second doubt. We have been assuming that gravity obeys Newton's law, which states that the force between two particles is inversely proportional to the square of the distance between them. But we might be mistaken about this. What if the correct law states that the force is inversely proportional to the *cube* of the distance between them? Can we relate this alternative possible law to a change that we should make in the deity's memento? Clearly not. It is possible that at a particular time two worlds could be precisely alike in where the particles are and what properties they have, even though they differ in laws. The memento is simply a record of where the particles are at a particular time, so it won't capture this distinction.

So in order to give an account of possibilities that does capture the difference between worlds that contain different laws, we need at a minimum to capture the different ways worlds can be over time. A simple way to capture that would be to take not just a single list, but a whole collection of lists, ordered in a temporal sequence. Each list would represent one moment in the complete history of this possible way that a world might be.

Even this may not be enough, because it could be that two worlds have identical histories of particle positions and properties, but differ in their laws.[5] So in addition to a sequence of lists, we may need the laws as well. A possibility, we might suggest, is a combination of a sequence of lists – effectively a history of the particle positions and states – and a list of the laws that govern that world.

But now consider the third doubt. If our tables all agree that the properties of particles are mass, charge, and position, then there seems to be no room for a table to represent possibilities in which particles have non-standard properties such as, for instance, charm, spin, and colour.[6] How could we rectify this?

Again, a fix does not seem too difficult to find. First, we need to allow that the number of columns in a table can vary. Second, we need to give labels to the columns, so as to identify the properties being used in those columns. So our account of possibilities will come to roughly this:

P_3 A possible way the world might be is a combination of three elements:
  i. A sequence of lists, each list being in the form of a table with N rows and M columns. Each row represents a particle and its properties.
  ii. A list of M column headings. Each column heading tells us what the properties are that are represented in the columns.
  iii. A set of laws.

Now, however, we come to the first and most fundamental doubt. What if matter is not particulate at all? Surely in that case, to use a structure like a list to represent a possibility is deeply misguided. A list requires that there is a countable number of basic entities, and that capturing the whole truth about the universe simply requires capturing the truth about these entities. But if there are no basic particles, then what sort of things do you put on the list?

---

5 This point is a matter of significant debate. A useful entry point into the literature is Carroll (1994). See also §32.
6 For the uninitiated, charm, spin, and colour are names coined by physicists for properties of particles discovered in the era of quantum mechanics, such as quarks.

Attempting to address this sort of concern will take us a long way into relatively dense and difficult debates in the metaphysics of possibility. I propose, instead, to withdraw from this debate. What has come to light, I trust, is that it is very difficult to give a completely general account of what possibilities are. The range of possibilities is surely very vast. No simple, easily systematisable way of representing these possibilities is obviously equal to the task of representing *all* of them.[7] Despite this disappointment for systematic metaphysics, however, for our purposes here we can still do interesting things with a way of representing a particular *subset* of the possibilities.

14 Creating a space of possibilities

Here is how we will restrict the space of possibilities, so as to make it more readily tractable. Let's assume the *same number of particles* as the actual world. Hence, if we were to represent these possibilities as lists, all lists would have the same number of rows as the deity's memento.

Second, we'll assume the laws are the same, in all possibilities. So there are no worlds where gravity attracts particles differently from how it does in the actual world. Third, we'll make the related assumption that the list of properties is fixed to be just those properties that are instantiated in the actual world. So there are no strange worlds where there are additional properties, or where an actual property such as mass is 'missing'. Fourth, we'll assume the static properties of each of the particles are fixed on what they actually are. That means that, in writing out each possibility as a list, we must always make the same entries under mass and charge, but we can make new entries under position and velocity.

These restrictions involve ignoring a great many ways the world could be. Infinitely many, no doubt. But that does not mean we will be short of possibilities by any means. There will still be infinitely many possibilities that satisfy this set of constraints.

Suppose we wished to *justify* focusing our attention on using this limited set of possibilities. Well, that justification would obviously be most effective

---

7 Even if we managed to get such a way of representing them, a host of deep questions about the nature of possibilities would remain. In my proposals P_1 to P_3, I have supposed that the possibilities just *are* the representations. That might be wrong, however. A representation of a possibility is arguably distinct from the possibility itself. That is presumably why we call it a *re*-presentation. But in that case, what could possibilities really be? Or is there something wrong in supposing possibilities to be real in the first place?

if we had a good account of what could be *achieved* by thinking about such possibilities, and that will be something I will look at later. But for now, before we look at what can be done with these possibilities, is it possible to say anything about what makes this set of particular interest? Is it possible to indicate why this set might be a justified restriction for some purposes? We can at least say this: these possibilities stand in a relatively *intimate* relation to the actual world. Because they contain precisely the same number of particles, instantiating the very same properties, and differing merely in position and velocity, these possibilities represent what we might call the 'mechanical possibilities' for this very world. By that I mean the possibilities represent different ways the very same world could have its constituents move over time. This is different from possibilities that involve having *different constituents*. To represent those sorts of possibility, we would need to allow changes in the number of particles, or changes in the properties of particles, or both.

So if there are contexts in which mechanical possibilities are especially salient, then that might afford us a justification for focusing exclusively on this set of possibilities. And I believe that, while there is certainly no conclusive justification of that sort available, I do think we can make it plausible that the mechanical possibilities are – for many contexts – a very good approximation to the relevant possibilities.

And it turns out that there are such contexts – but I will say more about that in the next chapter.

Having made this simplification of our task, by restricting the range of possibilities quite drastically, we can now develop an alternative method of representing the possibilities. This alternative method gives us a way of representing all the possibilities in a mathematical *space*.

A space, for a mathematician's purposes, is simply a set of points with some sort of structure. This structure involves the relations between points. It is in virtue of such structure that you draw inferences from two spatial relations to a third spatial relation. To take a trivially simple example (and ignoring some complications): Canberra is north of Melbourne. Sydney is north of Canberra. Therefore, Sydney is north of Melbourne. This sort of inference is supported by the structure of the space we live in.

Mathematical spaces, moreover, come in different varieties. For many years people thought that physical space conformed to the axioms of Euclidean geometry. But it turns out this is incorrect, and we now use Minkowski space-time to represent physical space (and time).

One of the most dramatic ways in which spaces can vary is in their dimensionality. We are already familiar from high school mathematics with the distinction between two-dimensional polygons and three-dimensional polyhedrons. But we can go further and study mathematically the nature of spaces with more than three dimensions. Spatially representing higher-dimensional spaces is difficult for creatures like us, because our representations can have no more than three spatial dimensions. But with sufficient abstractions, use of analogies, and so on, we can comprehend – in some sense – the possibility of additional dimensions, and mathematicians are able to reason rigorously about such spaces.[8]

Recall, then, that we have noted that in a list such as the deity's memento, there are numerous 'dimensions' along which we can make variations, thereby generating new possibilities. In representing the mechanical possibilities spatially, we are going to turn each dimension of variation into a dimension of a mathematical space. The space we get as a result is known as a state space. In particular, the sort of state space we are considering here is a *phase space*.

To get a working idea of a phase space, it will help if we start with an extremely simple example. Suppose we have a world that contains only *one* particle, in a *one-dimensional* space. That is, it is a single particle, in a universe that has the structure of a line. A deity's memento of such a world would contain only one row, and a number of columns. Some of the columns will represent static properties, like mass and charge. Since we are not interested in possibilities which change those properties, however, we shall ignore those columns. The remaining columns will be a single column for position in the single dimension of physical space, and a single column for velocity – again, in the single dimension of physical space.

In effect, then, this table has only one row and two columns of interest. Changing the values in the two columns will give rise to different possibilities.

In order to represent all the possible values that we could put under the position column, we can simply use a (mathematical) spatial dimension: the position of a point along that mathematical dimension represents the spatial position of the particle. So we use a dimension of a mathematical space to represent position in a dimension of physical space. (Remember that, in our simple example, physical space has only one dimension, but for real-world representations, we would need three dimensions.)

---

8 A charming attempt to assist the lay person to conceive of more than three spatial dimensions is Edwin Abbott's *Flatland* (2007 [1884]).

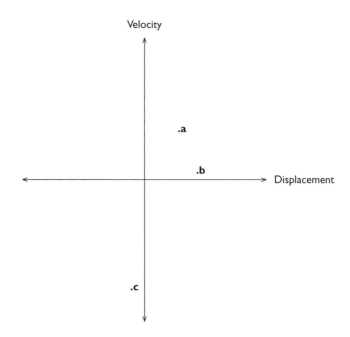

Figure 3.1 A two-dimensional phase space. The vertical dimension represents velocity, and the horizontal dimension represents position. Accordingly, the point **a** represents a particle somewhat removed from the origin, and moving at a moderate pace. The point **b** represents a slower moving particle, that is somewhat further removed from the origin than **a**. Point **c** represents a very fast moving particle – but moving in the opposite direction to the other cases. It is currently located close to the origin.

The other dimension of variation that we want to represent, then, is the velocity of the particle. We can apply the same trick as we did with position. We add a further dimension to phase space – our mathematical representation – and the position of a particle in this dimension represents velocity in physical space.

The result is a two-dimensional space which represents *all* the mechanical possibilities for our simple world of one particle in one dimension (see Figure 3.1). Each point in this space represents a way the world could be at a particular time, and each mechanical way the world could be is represented by some point in the space.

How would we go about extending this method to more complex possibilities? By adding an additional dimension to phase space for each additional property that can determine a possibility. So if we stayed with one particle, but looked at that particle in a two-dimensional space, we would need to add *two* extra dimensions: one to capture the possible positions of the particle in the second dimension of space, and a second dimension to capture the possible velocities of the particle in the second dimension of space. If we added a third spatial dimension, we would have to do the same again. With just one particle, every extra spatial dimension brings an extra two dimensions of variation for that particle: so we need to add two dimensions to our phase space.

What if we consider *additional particles*? Each particle can vary in its position and velocity quite independently of the other particles. So if one particle in two-dimensional space needs four dimensions in phase space, then a second particle will need *another four dimensions* to represent all its possible variations in position and velocity.

This may be hard to see, because you might be thinking that each particle is represented by a separate point in phase space. That would be wrong, however. *Each point in phase space represents the way an entire world is.* Therefore each point involves representing the way each and every particle in the world is. So when there are many particles involved, we need many dimensions of phase space, so that the position of a single point can represent the dynamic properties of all the particles.

As you have probably been able to work out, then, the total number of dimensions of phase space conforms to the following formula:

Dimensions of phase space =

2 × No. of particles × No. of dimensions of physical space

So for our world, which likely contains a very large number of particles, and at least three spatial dimensions, we will need six times a very large number of dimensions of phase space to represent the mechanical possibilities! Despite the unimaginable complexity of such a space for worlds like our own, the general strategy of employing phase space to represent possibilities is a useful one, because of the precisely understood mathematical properties of such spaces.

15 Possible histories

The deity's memento, then, amounts to something like a single point in a many-dimensional phase space.[9]

---

9 The match is not perfect because phase space does not explicitly represent the static properties of the particles. Moreover, phase space does represent velocity, which we initially suggested the deity might omit from the recording.

Recall that we are trying to devise an account of the possibilities that are compatible with our limited knowledge of the world. We have restricted the sorts of possibility we are considering so as to help us in constructing a powerful representational device for the possibilities: a phase space.[10] That restriction might have led us to ignore some interesting possibilities, but for now let us be optimistic that we still have all the important ones to hand.

But we do not merely wonder about different ways the world might be at the present time. We are also open to different ideas about how the past has gone, and how the future will go. Each point, recall, represents only how a world might be at a single moment in time. A set of points in phase space – being merely synchronic representations – seem inadequate to addressing these diachronic sorts of possibilities. I will call individual points of phase space 'snapshots', so as to stress the idea that they are representations of ways a world might be *at a single time*. Entire ways a world might be, over time, are possible 'histories'.

How could one go about representing a possible history in phase space? Representing a history would amount to representing a whole collection of instantaneous states, from the initial moment to the end of time, if such there be. Each of those instantaneous states can itself be represented by a point in phase space. So we can represent a possible history, quite simply, as a *collection* of points in phase space.

Moreover, we can expect these points to be related to one another by forming a continuous trajectory.[11] That is because of the nature of the laws that we use in the classical account. The world evolves, according to classical mechanics, in a continuous way from one point in phase space to another. It is never the case, for instance, that a particle moves in a discontinuous jump from one position in a given dimension to another. It must occupy all intermediate positions (in that dimension) along the way.

10 Arguably, we are restricting the set of possibilities that we are considering all the time. To use an example of David Lewis's, when I say that I cannot speak Finnish, I mean something like, in all worlds where my neural programming is much like it is at the moment, I do not speak Finnish. But in another sense, of course, I could speak Finnish: my vocal cords are in good order, and I have the ability to learn. So in at least some worlds where my vocal cords remain in good order and I retain the ability to learn, I do speak Finnish. In a less restricted set of possibilities, that is, there are cases where I speak Finnish. So relative to that restriction, it is possible for me to do it. See Lewis (1976a) for the original example, and Kratzer (1977) for more general discussion of the applications of restricted quantification over possibilities.

11 Intuitively, you might want to call it a *path*. The mathematician's term 'trajectory' merely registers the point that the intuitive idea of a path is somewhat unfamiliar in higher-dimensional space, and instead we need to rely upon a precisely defined mathematical term.

Recall that the laws, on the classical account, are deterministic. From the augmented memento, including velocity information, all future and past states of the world are determined. A point in phase space, moreover, contains the same information as the augmented memento. So a point in phase space, plus the laws, determines the entire world history, which is represented by a trajectory through phase space.[12]

So for each point in phase space, if the deity cared to perform the necessary calculations (such calculations are utterly beyond *our* ability for systems of more than a handful of particles), it could identify the trajectory that is determined by each point. If you visualise a three-dimensional space with a number of wiggly paths criss-crossing through it, you get some sense of what this would be like – though in a hugely simplified way. The visualisation is simple in two obvious respects. First, there are not sufficiently many dimensions. Second, you are almost certainly not imagining sufficiently many trajectories. Plausibly, there are infinitely many possible histories, so there should be infinitely many wiggly paths criss-crossing your visualisation.

Note that, because the laws are deterministic, the trajectories will never intersect. Another way of putting this idea is that, if two putatively distinct trajectories overlap in some place, they overlap everywhere, and therefore are not distinct at all. This is because the laws are fully deterministic in both directions. Any state in a world history determines the entire future and entire past. Two distinct trajectories coming together and meeting at a point *x* would indicate that point *x* determines two different possible pasts, or two different possible futures. And that simply cannot be, given determinism. So when visualising the tangle of paths in our crude image of phase space, the paths should be like spaghetti. They can get very close to each other, but never strictly *coincide*.[13]

---

12 Also recall that in Chapter 2 I stressed the ambition to *completeness* of the classical account. All there is to world history, according to this view, is a collection of particles moving in space. Representing world history as a trajectory in phase space, then, is simply a corollary of this ambition.

13 A subtlety I have overlooked here is that you might think it possible for a world simply to begin, *ex nihilo*, at any point in phase space. If that is so, there could be multiple ways that a world might be which overlap in phase space: we could have one world which begins at what we will call 't_1' and runs through to 't_∞'. A second way the world might be is to spontaneously begin as a perfect duplicate of the first world at t_2, and then run through to t_∞. So these two ways a world could be are distinct, though one entirely overlaps the other in phase space. I'll ignore such possibilities henceforth.

4 | Possibilities of thought

In the previous chapter we examined phase space as a way of representing mechanical possibilities. These are the ways the world might be, assuming something like the classical picture is correct, and assuming that the world contains a certain number of particles.

Phase space does not capture *all* of the possibilities, but it is – I suggest – a useful set of possibilities for certain purposes. In this chapter, I want to focus on the limitations of phase space, and similarly constructed spaces of possibilities. Knowing the limits of these approaches, we'll be better placed to use them with confidence in an analysis of chance.

16 Propositions in phase space

Typically, when we entertain thoughts about the world, it is not in the same degree of detail as the highly specific possibilities that were introduced in the previous chapter. For instance, we might wonder about whether or not there is an elephant in the room. If we try to identify, in terms of mechanical possibilities, which mechanical possibility correlates with an elephant being in the room, then we find that there is a mismatch between the specificity of our thoughts and the mechanical possibilities. There are *lots* of different mechanical ways you can arrange particles in space so as to have an elephant in the room. So what might have seemed like 'one' possibility turns out to correspond to many different mechanical possibilities.

What we want to represent is the content that can be embodied in a thought, a belief, or a sentence: a *proposition*, as I shall call it. As we have seen above, the proposition that there is an elephant in the room lacks the highly determinate content which we would expect to be present in *the world* if there really were an elephant in the room. A world with an elephant would involve a very large number of particles, each with determinate properties of mass, charge, position, and the like, and those particles would constitute the elephant. The proposition that there is an elephant in the room might commit us to there being *some particles*, located *somewhere* in the vicinity of the room, but does not include the idea that the particles have any particular properties.

One way to think of what the proposition does commit us to, then, is to think of all the ways it might be true, and to identify something *in common* between all of those ways. So suppose we could go through phase space, picking out each point (each snapshot) that represents a way the world could be such that it is true that there is an elephant in the room. (To resolve possible ambiguities, I should specify a date and time, and precisely what room I mean. I'll ignore that complication here.) I then form the set of all such snapshots – call it E.

What property do all these snapshots have in common that no other snapshots have? Quite simply, all and only these snapshots are members of E. So if we want to say in terms of phase space what the content of this proposition is, we cannot use a single point to do that because it has too much specific structure. By forming the set E, however, we have an entity that averages out the idiosyncratic structure of its individual members, and the property of *being a member of E* captures just the right amount of structure to get at the idea of there being an elephant in the room, but no more.

Suppose we do the same thing with another proposition: 'There is a mouse in the room.' We form a set, M, of snapshots in which this is true. Some of the snapshots in this set will represent worlds in which there are three mice in the room (if there are three mice, of course it is true that there is one). Some of the snapshots will be ones where the mouse is alive, some where it is dead. Some will even include other animals in the room – even elephants.

Having identified sets for some sentences, we can perform simple operations to get logical compounds of those sentences. So if we entertained the disjunction: 'Either there is an elephant in the room, or there is a mouse in the room, or both', we should form the set which represents the structure of that proposition simply by taking the set-theoretic *union* of E and M. If we were interested in the conjunction: 'There is an elephant in the room and there is a mouse in the room', we should form the relevant set by taking the set-theoretic *intersection* of E and M. (See Figure 4.1.)

17 Troublesome thoughts

This method has much to recommend it. By set-theoretical operations on points of phase space ('snapshots'), we can effectively dispose of the

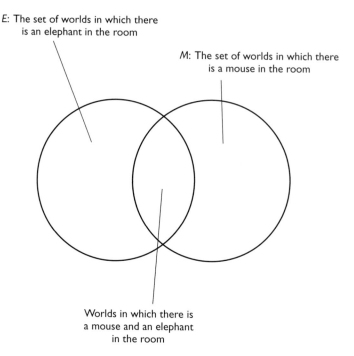

Figure 4.1 Sets of snapshots that might be used to represent propositions. Inside the set *E* are all the possible ways that there might be an elephant in the room.

unwanted structure in individual snapshots, and thereby get entities with the right sort of structure to represent thoughts and propositions.

There are at least two important species of thought, however, which seem to defy incorporation into this framework. The first is what we might call *impossible* thought. Consider someone who attempts to entertain the proposition 'There is a greatest prime number.' This thought is false.[1] Not only that, but it is *necessarily* false. So the set of snapshots in which this proposition is true is the empty set. There is seemingly no structure to this thought at all.

That might not worry you. The thought has no true structure, you might think, because the thinker is so terribly *confused*. Tempting though that thought is, it fails to capture the distinctive role that this thought has in our

---

1 The proof is so simple, it bears repeating for the few who have not seen it. Take any number you allege to be the largest prime, *k*. Take all the prime numbers less than or equal to *k*. Multiply all these numbers together, and add one. The resulting number cannot be divided by *k*, nor by any of the primes smaller than *k* – for if you were to divide by any of those numbers, the remainder would be one. Therefore it is either itself prime, or it is divisible by a prime number greater than *k*. In either case, there is a prime greater than *k*, thus refuting our initial supposition that there is a greatest prime.

reasoning as opposed to other counterlogical claims that we might make. Consider another impossible thought: '2 + 2 = 5'. Someone who believes this falsehood surely believes something different from the person who believes that there is a greatest prime number. But by the method we have described to identify the contents of thoughts, this second claim has precisely the same content as the first: the empty set.

For some purposes, it may not matter if we treat all impossible thoughts in the same fashion. But clearly to say that '2 + 2 = 5' and 'There is a greatest prime' have precisely the same content is not capturing the content of a belief in the sense of something that actually features in the psychology of believers. So for this sort of purpose at least, our model is not going to be useful. That does not condemn the model outright, for it might still have other uses, but we must concede that the model seems inadequate for this purpose.

The second sort of worry is also about thoughts that are true nowhere in our space of possibilities, but not because they are logically incoherent or metaphysically impossible. Think of an early Greek philosopher who has developed a basic theory of physics. This philosopher has beliefs like the following:

- Water is an element.
- Gravity's sole effect is to attract all massive objects to the centre of the Earth.
- Air is infinitely divisible.

These thoughts are false in all of the points of phase space. Water is not an element – a basic, non-decomposable substance. Rather it is a chemical compound of more basic elements. And this is the nature of water throughout phase space. Gravity does attract massive objects to the centre of the Earth, but that is not its sole effect. Gravity is not a force peculiar to the Earth: it operates between massive bodies at large. Because all the worlds of phase space are governed by the same laws, there is no world where the Greek philosopher's belief about gravity is true. Finally, air is not infinitely divisible, but particulate. And this is false in all snapshots because in all points of phase space all matter is particulate.

Moreover, by focusing on impossible thoughts, or on propositions relating to things that are true in no points of phase space, I have been focusing our attention entirely at one end of the problem. But the same problem also affects thoughts that are *true* in every point of phase space. Thoughts like 'All matter is particulate', '2 + 2 = 4', and 'Masses attract each other' are true in every snapshot, so their content – understood in terms of phase space – is the same: the set of *all* points. Clearly, however, these thoughts are very different in what role they play in our cognitive lives.

The second type of problematic propositions – including those which are not strictly impossible or necessary, but are nonetheless false in all points of phase space or true in all points of phase space – arises from the artificial restrictions we made to the space of possibilities in the previous chapter. In order to have a well-regimented phase space that could be given a precise definition, we needed to make various assumptions about the range of possibilities. This meant that the only possibilities we considered were the possible arrangements in space and time of *a particular set of particles, subject to a given set of laws*. Possibilities that involve *other substances, other particles*, and *other laws* were omitted. Unfortunately, however, we evidently can and do have beliefs about such possibilities. So again, we cannot use phase space to model the contents of such beliefs.

There is something even more troubling about this second problem for phase space as a means of representing the contents of belief. Recall that the first problem related to thoughts that were impossible. We might be interested in studying the content of such beliefs for the purposes of linguistics or psychology, but we are unlikely to need to study them for the purposes of physics or metaphysics. Impossible thoughts surely do not correspond to a way the world might be, or actually is. So to the extent we want to study the way the world is or might be, we do not need to worry about impossible cases. Evidently, however, we cannot be so dismissive about the possibilities that our Greek philosopher was contemplating. These are all ways the world really could have turned out to be. Of course, we now think it overwhelmingly unlikely that the world conforms to any of these beliefs, but we had to actually go out and do some scientific inquiry to eliminate those possibilities. We cannot say – as we can in the case of false beliefs in mathematics – that we simply needed to think a bit harder, and we would have realised it was never really possible for two and two to equal five.

In order to represent the contents of thought, we would do better to make two changes of approach. First, in order to accommodate impossible thoughts, we need to model the content of propositions using elements that are not themselves worlds, nor world-like things, such as points in phase space. The reason is that these sorts of entity have a highly refined structure – a structure that is capable of capturing all the detail of the way the world is or might be without falling into logical inconsistency.

Thoughts, in contrast, are much more like language than like the world. Sentences can be ungrammatical, yet still convey some meaning. Sentences can be self-referential, and involve paradox, wit, and wordplay. Sentences can represent the world, but they can all too easily fall into contradiction whenever we try to represent something very complex. Thoughts, too, are vulnerable to such dangers. So the contents of thoughts and of sentences – the things we have been trying to represent in phase space – clearly can be far more 'messy' in their structure than can the world. An adequate account of the contents of thoughts, then, must allow for this sort of mess.

Second, in order to accommodate more remote possibilities, such as those entertained by the physicists of the past (not to mention those to be entertained by physicists of the future), we will need to liberalise our account of possibility. In particular, we might need to abandon the hope that there is a space of possibilities with a *finite number of dimensions*. We are then left with two alternatives, neither very appealing. Perhaps there is an infinite-dimensional space of possibilities of which phase space is merely a cross-section. Or perhaps the realm of possibility is simply too perverse to admit of any systematic structure.

And note that even if we could give a decent account of these more remote possibilities, we would still have the problem of incoherent or metaphysically impossible thoughts. So addressing the second problem – at least in the most obvious fashion – does nothing to address the first.

I have no such augmented account of possibilities to offer here, in order to address either the first or the second problems. My principal aim is to show that useful work can be done with phase space and related accounts of possibilities. The purpose of the foregoing is to chasten the ambition of others who would try to use phase space for representational purposes for which it is unfit. That said, phase space can be used to represent an important subset of our beliefs about the world. And this use of phase space is – it will turn out – of genuine value to us as agents who want to know things.

18 Counterfactual possibility

Thus far, I have been focusing on ways the world might *be*, given some of what we know. But there is another sense of possibility that has been given a lot of attention by philosophers: there are ways that the world is *not*, but *might have been*.

The early Greek philosopher who thought that gravity was a force which attracted everything in the universe towards the centre of the Earth was presumably not thinking something impossible. This is a way things could have been. A more mundane way in which we might want to think about possibilities that could have been, but are not, is when we entertain hypothetical thoughts. Consider: 'Had I set the alarm last night, I would have got out of bed twenty minutes earlier.' As it happens, I know that I did not set the alarm last night. This is not a way the world *might be*. But it is a way things *could have been*.

Conditional sentences of this sort are known as counterfactuals. We might have hoped that phase space would be good for modelling counterfactuals.[2] But the idea runs into some serious difficulties. The central problem is roughly this: suppose I ask you to entertain a possibility which we know did not obtain: 'Had you jumped out of your bedroom window this morning...' Very plausibly, if you had done this, you would have injured yourself because – say – your bedroom is on the second floor. What sorts of histories in phase space would represent worlds where this event – your jumping out of the window – occurs? Because the histories of phase space are deterministic, they are going to have to be histories where the past was different in some way. We need to find some way of tinkering with the conditions of yesterday, such that you would have been deterministically caused to jump out of your bedroom window today.

So imagine you find a history where things were just a little bit different yesterday. Now ask: what would this world have been like the day before? Again, it has to be different from what actually happened the day before yesterday, because otherwise things would have been deterministically caused to stay the same as they actually did. So we can conclude that, had you jumped out of the window this morning, things would have been different yesterday, and the day before.

And so forth. By the same reasoning, we can conclude: 'Had you jumped out of the window this morning, the entirety of world history would have been different.'[3] This is a silly conclusion, and it is very hard to see how it could be avoided.

That's not to say that counterfactual thoughts and sentences should be shunned. But we must at the very least invoke a different conception of possibilities if we are to model counterfactual thoughts. We probably cannot settle just for the mechanical possibilities – otherwise we are condemned to problematic conclusions like the above.

---

2 I discuss this idea further in Chapter 11.

3 Trying to eliminate this conclusion from a theory of counterfactual semantics, without inserting a time asymmetry by brute stipulation, has proven very hard. See, e.g. Elga (2001); Lewis (1979b); and Price (1996: chap. 6).

19 Macroscopic states

In our ordinary practice, we do not study physical systems at the level of particles. Even a very simple system, like a box of hot gas, is not studied by measuring the position and velocity of individual particles. Rather, the sort of measurements we have found useful to make on such a system record the *temperature* of the gas, the *pressure* it exerts on the walls, and the *volume* of the box. None of these are properties of individual particles.

Even so, we can be confident that the properties of the gas in the box are related to the properties of the constituent particles. More than that, we can be confident that the properties of the gas *depend* upon the constituent particles. Change the properties of the particles in the right way, and you will change the properties of the gas. Moreover, you *cannot* change the properties of the gas without changing the properties of the particles.

Not every change in the properties of the particles, however, will change the properties of the gas. This is because the properties of the gas are 'multiply realisable'. Different arrangements of particles can realise the very same properties of the gas. Multiple realisability occurs in a wide variety of phenomena. We have already noted the possibility that the property of *being an elephant* was multiply realisable, because there were many different arrangements of particles that would satisfy that property. Here, the phenomenon is much easier to analyse than a complex case like elephanthood, because the properties in question are now understood by physicists to be *statistical averages* of the behaviour of individual particles.[4] Two populations of humans, for instance, can have the same average height, even though no member of the first population is precisely the same height as anyone in the second population. Precisely the same sort of phenomenon is at work here, with properties like pressure and temperature.

So for a gas at 100°C, suppose that we focus on just two of the particles in that gas: particles A and B. Temperature is a function of the statistical average of momentum squared. We will suppose that the momentum of A is 3 units, and that B has a momentum of 4 units. The average squared momentum is then:

$$\frac{3^2 + 4^2}{2} = 12.5$$

---

4 Earlier, physicists defined temperature and pressure 'phenomenologically' – that is, in terms of measurements that could be made and felt at the level of human sensation. The reduction of these properties to such simple statistical averages of mechanical properties remains a rare example of straightforward success for scientific reduction.

Various different configurations of these particles will be ways in which the gas has the same temperature. One is a configuration in which all the other particles remained the same, but A and B swapped momenta. Another would be one in which A had zero momentum and B had a momentum of 5 units. Indeed, keeping the other particles fixed, any state of A and B, such that the sum of their squared momenta is 25, will be one in which the temperature remains $100°$C.

Again, there will be infinitely many such possibilities, because there are infinitely many ways for two positive real-valued variables to reach a constant sum. And the same story can be told for all the particles in the gas. As long as the sum of all their squared momenta remains constant, you can assign any momentum you like to any particular particle, and the temperature will remain unchanged.

Physicists call the total condition of a system, specified in terms of properties like temperature, pressure, and so forth, the *macro-condition* of the system. Macro-conditions can be usefully represented in phase space. Given the multiple realisability of macro-conditions, we cannot represent a given macro-condition using a single point in phase space. However, in much the same way we could represent the thought that there is an elephant in the room using a *set* of points, we can also represent a macro-condition using a set. For reasons we need not go into here, the points in this set will not be scattered higgledy-piggledy through phase space. Rather, the set of points will form something like a collection of blobs in phase space: one or more solid regions, each of some complex shape.

There is an attractive simplicity to the contrast, then, between the complete *micro-condition* of the world – the sort of thing recorded in our imaginary deity's memento; and a *macro-condition* of the world – the sort of thing we could in principle measure.[5] The micro-condition is equivalent to a point in phase space. The macro-condition is equivalent to a solid region of phase space.

Why is it that phase space can be used to model a macro-condition of the world, but not beliefs about the world?

It is because phase space is a very limited set of possibilities. The range of possible beliefs about the world vastly outstrips what can be represented in phase space. As I have already indicated, it is possible to have contradictory

---

5 We can actually measure – to a high degree of accuracy – the macro-condition of small, simple subsystems of the world. For a system as large as the entire world, however, it is a fantasy to suppose we could ever have an accurate measurement of its entire macro-condition.

beliefs about the world and it is possible to have beliefs about the constitution of the world which violate the assumptions of phase space. So beliefs in general cannot be captured in phase space.

The reason macro-conditions can be captured is because they do not present such problems. At least, there certainly are no such things as 'contradictory possible macro-conditions'. Of that we can be quite confident. It might be that there are possible macro-conditions of the world which defy the assumptions we make in constructing a phase space. But for current purposes, we are assuming that the world conforms to the classical account, so we can set that concern aside.

20 Phase space and epistemic possibility

The classical picture of the world is false. But even though it is not strictly correct, it is – for many purposes – an excellent method of organising our knowledge about the world, and using that knowledge to make further predictions. Practising physicists still use classical physics for many purposes, either because the additional complications of later physics will leave them unable to perform the calculations they need or because the additional accuracy that could be obtained by doing it otherwise is not worth the effort.

We also know that we can, for limited subsystems of the world, come to have relatively accurate beliefs about the macrostates of such systems. We can measure temperatures, pressures, and so forth, to very high degrees of accuracy. Moreover, we cannot, for physical systems of this type, ever hope to have similarly detailed knowledge of their *microphysical* state. Having a very good knowledge of the macrostate is close to the limit of what we can know about such systems.

So phase space – which embodies the physical assumptions of the classical picture, and which is a good way of representing macrostates – is a good approximation of what is possible, *relative to the available evidence*. Perhaps phase space, then, can be used in an account of chance, which I have characterised as roughly the *best degree of belief, given the available evidence*.

5 | Chance in phase space

Phase space is a good model of what we are interested in when we talk about a certain range of possibilities. Not all possibilities can be captured in phase space, but, for certain purposes, many of the interesting ones can be. In this chapter, I will introduce the idea that we have not yet seen the most powerful and useful application of phase space: to model certain types of probabilities or chances.

21 The leaking tyre

You wake up to find yourself in a closed room, isolated from outside causal influence. In the middle of the room is a bicycle tyre. Because all else is quiet, you can hear that the tyre is hissing very gently, and as you go over to it, you are able to locate the small stream of air that is leaking out of a tiny hole, producing the noise. You judge, by a squeeze of the tyre, that it will take some time before the air will stop hissing out of the tyre.

What is the macro-condition of the room right now? It is one in which there is a large volume of gas at relatively low pressure, and a small volume of gas in the tyre, at relatively high pressure. Moreover, there is a small aperture between these two volumes of gas.

What, then, is going to happen? Obviously, the air will continue to leak out of the tyre until the two volumes of gas are at the same pressure.

Now here is a curious thing we have already noted about the classical account. The laws of the classical account are invariant under time-reversal. The remarkable upshot of this is that, for any process that can happen in one direction in time, it can also happen in the other direction. Therefore, though we are highly confident that the air will leak out of the tyre, there is a possible process, compatible with the laws, whereby the perfect reverse of this occurs: the particles of air rush *into* the tyre from the room, creating an ever greater difference between the pressure in the tyre and the pressure in the room. This process is so bizarre that – no matter how confident physicists are in the mechanical laws – many of us would surely find it impossible to believe that such a thing could happen.

Moreover, you might worry, wouldn't the reverse process be in conflict with the laws of thermodynamics? That thought is correct. The laws of thermodynamics state that closed systems always move towards a condition of equilibrium. For the system I have described, the equilibrium condition is one in which the pressure is the same throughout the room and within the tyre. So if it is possible for the air in the room to spontaneously inflate the tyre, then the so-called 'laws' of thermodynamics cannot be genuine laws. At best, they might be true generalisations: they tell us what actually happens, but they do not tell us what *can* and *cannot* happen.

You might hope that the way out of this impasse is to reject the classical account, with its time-symmetric laws, and to embrace some more sophisticated physics which contains an inherent asymmetry with respect to time. It so happens that physicists have found evidence that some fundamental physical processes are temporally asymmetric or time-irreversible. But it also seems unlikely, most believe, that these types of time-irreversibility will explain the irreversibility of thermodynamic processes.[1]

The way physicists have reconciled this issue is to adjust their conception of the 'laws' of thermodynamics. These laws are not strict and deterministic laws, but are in some sense probabilistic generalisations. So the laws of thermodynamics do not entail that it is *impossible* for the air to rush back into the tyre and re-inflate it. Rather, they should be understood to state that this is *enormously unlikely*.[2]

But even this interpretation has difficulties. Given that we are assuming the classical account, where do the probabilities come from? The mechanical laws in classical mechanics are strictly deterministic. Either the particles of air will rush back into the tyre, or they won't. It does not look as though there is room for any objective probability claims about this matter.

22 Counting possibilities

The macro-condition of the room, just after I wake up, is one of high pressure in the tyre and low pressure in the room. One way of representing

1 For a variety of views on the significance of these processes for the broader question of whether there is a fundamental time asymmetry in the physical structure of the universe, see Horwich (1987: 55–6); Price (1996: 18); and Maudlin (2007a: 117–8, 135–7).

2 See Albert (2000: chap. 3) for a more thorough discussion of the physics involved. In particular, although the explanation I give here suffices to explain why it is very unlikely, in the future, that we will see air rushing into the tyre, it remains unexplained why it is not also unlikely that in the *past* the air rushed into the tyre. I will return to this issue in Chapter 11.

this macro-condition is to draw a solid blob (or possibly a collection of blobs) in phase space. While I personally am in no position literally to *draw* such a blob, we can study such entities in phase space using mathematical techniques. This is because of the well-defined nature of phase space and the way we understand macroscopic properties of a system in terms of statistical functions of the properties of individual particles.

Each and every point in the blob is a possible micro-condition that the room could be in, compatible with the macroscopic properties which the room has. Some of these micro-conditions will be ones in which the future behaviour of the system is perfectly normal: the air will continue to leak out of the tyre until equilibrium is reached. Some of these micro-conditions will be bizarre ones in which the future behaviour of the system is anti-thermodynamic. For instance, the air could rush back into the tyre and increase the pressure difference between the tyre and the rest of the room. Alternatively, other anti-thermodynamic behaviour could occur. Some of the nitrogen molecules in the air could dramatically slow down and turn into a puddle of freezing cold liquid nitrogen on the floor. Or all of the air molecules could cluster in a corner of the room, creating a near vacuum in the rest of space. We do not know which of these micro-conditions we are in, but we have good reason to be confident that it is *overwhelmingly likely* that we are in a 'normal' micro-condition, and that none of these bizarre anti-thermodynamic processes will occur.

What makes it the case that the normal micro-conditions are more likely than the bizarre ones? An intuitively appealing answer is that the normal micro-conditions simply *outnumber* the bizarre ones. Could it be that if we counted up all the normal micro-conditions, and also all the bizarre ones, and compared them, we would find that the sheer number of micro-conditions that lead to normal behaviour is vastly greater than the number of micro-conditions that lead to bizarre behaviour? There is significant intuitive appeal, then, to the thought that it is going to be much more *probable* that we are in a normal micro-condition, rather than a bizarre one. If all goes well, perhaps we would be able to calculate the probability according to a simple ratio formula:

$$\text{Probability(normal)} = \frac{\text{No.(normal)}}{\text{No.(normal)} + \text{No.(bizarre)}}$$

Delightfully simple though this suggestion is, it is hopelessly inadequate. The problem is that simple ratios are mathematically feeble in the face of the vast number of possibilities we are dealing with. If we were to substitute in the appropriate values to the above equation, we would find that the probability would be equal to infinity divided by infinity. And any such ratio is undefined. It has no meaningful value.

But this is not cause for despair; it is simply cause to invest in more powerful mathematical techniques. It so happens that we already have, and make frequent use of, mathematical techniques to assign finite numbers to sets which contain infinitely many members. And we do this so that we can make meaningful comparisons of *size*, in some sense, between such sets.

To illustrate, consider how you might go about answering the question: which of two balls is larger? Suppose one is a golf ball, and the other a basketball. Both of these are objects in three-dimensional space. You might think, then, that the way to work out their sizes is to count the *number of spatial points* which each ball occupies. That is indeed one way of assigning 'sizes' to the balls, but it is not a way that tracks what we are interested in when we try to talk about size. Because there are infinitely many points in any extended region, we would get the bemusing answer, using this method, that the balls are precisely the same size. Each contains infinitely many points. Indeed, this technique would yield the answer that *all extended objects are the same size!* Clearly, we need some more sophisticated mathematics than mere counting.

23 Measuring volumes in phase space

The mathematical technique used to assign a finite 'size' to an infinite set is called measure theory. Measure theory allows us to attribute *length* to lines, *area* to surfaces, and *volume* to solids. In all such cases, there is a set with infinitely many members – points in the line, points on the surface, and points in the solid – and a measure function returns a finite value for each of these sets, allowing us to make useful comparisons between different lines, shapes, and solids.

So in similar vein, we can apply a measure to a solid in phase space, and use this to calculate a 'volume'.[3] Instead of comparing the *numbers* of processes that are bizarre or normal, we compare the *volume* that such processes occupy in phase space. If all goes well, we could then analyse the probability of a normal process as the ratio of the volume of normal

---

3 This is not necessarily the same measure of volume that we use in normal, physical space – but it is sufficiently analogous that it is helpful to use that term to describe what we are doing.

processes to the volume of all processes:

$$(1) \quad \text{Probability(normal)} = \frac{\text{volume(normal)}}{\text{volume(normal)} + \text{volume(bizarre)}}$$

The immediate problem with this is that there are *many* possible measure functions. There are, in fact, *infinitely many* such functions. And they assign different measures to the same regions of phase space. Which one of the measures is the true measure of volume – or at least the true measure that we should use for determining the chance of a process?

One thing we can do is to restrict our attention to measures that obey the formal axioms of probability. Measures that are not even mathematically like probability are obviously of no interest. But that is still going to leave us with infinitely many candidates.[4]

You might hope that there are some measures which are more 'natural' than others. There might even be one that simply stands out as the obvious choice. Any other way of assigning a measure to a set, you might think, just looks weird, or flouts what we know about the physics, and can be ignored on those grounds alone.

There is one measure, known as the *Lebesgue* measure, which does not flout any of the physical constraints we might like to impose on the choice of measure, and which is in some sense 'uniform'. It treats all dimensions of phase space *equally*, in a manner of speaking. But even so, this is scant reason to be confident that it is the correct measure. Can we really be so confident that the correct measure is uniform? Recall that the different dimensions of phase space track completely different properties. Half of the dimensions relate to velocity, and half relate to position. Why should we expect these dimensions to contribute to the correct measure in the same way? These questions should make us very uncomfortable about boldly picking a measure without any independent basis to verify that choice.

As it happens, the Lebesgue measure works quite well. That is, if you use it in the formula (1) given above, it will duly tell you that, given the current macro-condition, the probability that you are in a bizarre micro-condition which will give rise to anti-thermodynamic behaviour – such as air spontaneously moving into a tyre, causing it to inflate – is spectacularly

---

4 It will include, for instance, silly measures which attribute measure one to the proposition 'Talking poached eggs exist' and everything entailed by that proposition. That measure would be an apt model of the state of mind of someone who dogmatically believes in the existence of poached eggs that can talk. But that is clearly not a very useful measure for understanding objective chance!

low. And that prediction fits with our experience. Such processes do indeed appear to be extraordinarily unlikely.

Consquently, physicists have simply hypothesised that the probability of a particular thermodynamic behaviour is given by the relative Lebesgue measure, in phase space, of micro-conditions which lead to that behaviour. This hypothesis is known as the *statistical postulate*. Because this hypothesis has produced predictions that fit our experience well, we can tentatively conclude that it is correct, or close to correct. That is, we can conclude that the Lebesgue measure really is getting at the relevant property of a set of possible processes to give us the probability of such processes.

This concludes my attempt to introduce readers who know very little about physics to classical statistical mechanics, which is the most important example of chance-like probabilities in classical physics. What is tempting, in light of this picture, is to see if we can generalise it to provide a theory of what chance is, in general. That is the topic of the next chapter.

6 | Possibilist theories of chance

> We understand the actual world only when we can locate it accurately in logical space.
>
> (Bigelow and Pargetter 1990)

24 Possibilism

In the opening chapter, I characterised chance as the degree of belief recommended by the best identifiable advice function, given the available evidence. In the following chapters, I have sketched the sort of conceptual tools used by physicists to obtain probabilities in classical statistical mechanics: a measure over phase space. So a naive response to this presentation is to think that classical statistical mechanics actually tells us *what makes a fact of chance*. It is a *measure* over a space of *possibilities*. The space of possibilities contains the states the system might be in, given the available evidence. Given the right measure – something that we have confirmed by experience – chances are just facts about the relative measures of different macro-conditions. Call this the *modal volume theory* of chance.[1] In order to assess this idea adequately, we first need to unpack it.

Chances are ratios of volumes

Given I toss a coin, what is the chance that it will land heads? The answer has something to do with two sets of possibilities: that in which I toss a coin, and that in which I toss a coin *and* it lands heads. According to the modal volume theory, it is something to do with the relative 'volume' of these sets. (Of course, we do not mean volume in the same sense that we refer to volume of physical space: but the sort of measure we use will be analogous to our familiar measure of volume, so it is a homely term to use.)

---

1 My presentation of this idea draws from an unpublished – and abandoned – paper co-written by myself and John Bigelow. Indeed, the proposal that probabilities might be ratios of volumes of possibilities is original to Bigelow. The seminal papers in which Bigelow introduced these ideas are Bigelow (1976, 1977).

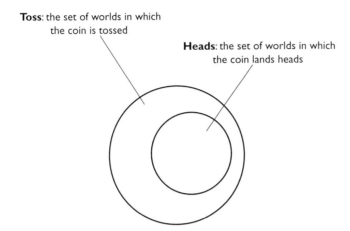

Toss: the set of worlds in which the coin is tossed

Heads: the set of worlds in which the coin lands heads

Figure 6.1 The conditional chance that, given I toss the coin, it lands heads, is the ratio of the volumes of **Heads** to **Toss**.

The possibilities in which the coin is not only tossed, but also lands heads, are clearly a subset of the possibilities in which the coin is tossed. So the volume of the set of possibilities in which the coin lands heads will be no larger than the volume of the set in which the coin is tossed. So the ratio of the volume of the lands-heads possibilities to the volume of the coin-toss possibilities will be less than or equal to one. The probability that we are interested in is just this ratio. (See Figure 6.1.)

Putting this more formally, for any conditional chance of the form:

$$\text{chance}(p, \text{given } q)$$

– that is, the chance that p, given q – the value of this function is equal to *the ratio of the volumes of two sets*. The first of these sets is the set of possibilities in which 'p and q' is true, and the second is the set of possibilities in which 'q' is true. Take the volumes of both these sets and find the ratio of these two volumes. That is the chance.

$$\text{M}_1 \qquad \text{chance}(p, \text{given } q) = \frac{\text{volume}(p \cap q)}{\text{volume}(q)}$$

There are at least two immediately obvious problems with M_1. First, by defining conditional chance as a ratio, it is essential that the volume of the condition does not equal zero. If it does, then the ratio is undefined.

Using the best available measure over phase space, events of a type with measure zero really can happen. If we think that there can be chances that are conditional on such events, then M_1 will fail as an account of chance. For example, suppose you throw a dart at a dartboard, and it lands precisely on

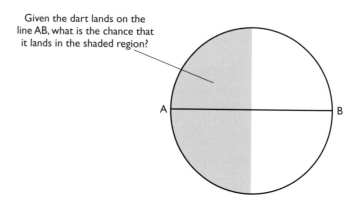

Given the dart lands on the line AB, what is the chance that it lands in the shaded region?

A   B

Figure 6.2 A question about chance which looks troublesome, on the modal volume theory.

the line that runs between the leftmost extreme and the rightmost extreme. Given that, what is the chance that it lands on the left-hand side of the dartboard? (See Figure 6.2.)

Such a query seems like a perfectly sensible question about chance, but the volume of phase space in which the dart lands on precisely that line has measure zero. So the chance is undefined, according to M_1.[2]

Second, to have a well-defined chance that p, given q, it will be necessary that both the terms on the right-hand side of the formula are well defined. They cannot be vague or indeterminate, for instance.

Many of the matters in which we are interested, however, are not the sorts of proposition that can be modelled in a limited space of possibilities such as phase space (for the sorts of reasons discussed in §17). Because our everyday language and thought is infected by vagueness, and because our ordinary language and thought is compatible with all sorts of bizarre possibilities that are left out of phase space, there is no set in phase space which is identical in content to an ordinary proposition like 'the coin lands heads'. And in that case, there is no particular volume of phase space that can be ascribed to an ordinary proposition like 'the coin lands heads'.

---

2 Alan Hájek (2003b) has raised similar concerns as difficulties for any attempt to *analyse* conditional probability in terms of a ratio of unconditional probabilities. Hájek advocates adopting conditional probabilities as conceptually *primitive*. So Kolmogorov's ratio formula:

$$\text{Ratio.} \qquad \text{prob}(p, \text{given } q) = \frac{\text{prob}(p \text{ and } q)}{\text{prob}(q)}, \text{ if prob}(q) > 0$$

which was put forth as a *definition* of conditional probabilities, is instead to be understood as a *constraint* that conditional probabilities must meet.

So the modal volume theory of chance appears to be unable to give us chances for paradigmatic chance events, such as coin tosses. This looks like a very serious limitation.

There are a number of responses which the defender of the modal theory of chance might make at this point.

With respect to propositions with zero volume, the best response is probably simply to concede that this limits the scope of the theory, but to deny that it is a problem. These events are in some sense philosophers' constructions, and they do not correspond at all well with chance claims that we actually need to make in physical theorising: perhaps it is therefore reasonable for the modal volume theorist to advocate a theory which is silent about such alleged chances. The resulting thought would be that chances are only well defined for claims about the world being in a particular *macro-condition*. Claims about the world being in a particular micro-condition have measure zero, so any chance claim conditional on a micro-condition will be undefined. But we cannot possibly identify what micro-condition the world is in. So it may not be a great cost to forgo chance claims conditional on particular micro-conditions.

Further, recall that I have suggested that we characterise chance as the *best* thing to believe, given the available evidence. To deny that there are any chances conditional on measure zero events is to deny that there is a best thing to believe about such conditions. Given we have very poor epistemic access to such unlikely events, perhaps there is no best thing to believe about them. This reply looks somewhat desperate, however, given that we can confidently hypothesise that if we *did* receive evidence that, despite having measure zero, *entails* the proposition we are interested in, then it seems obvious that the chance is one. Even if we never are in this position, the modal volume theory should not write it off as seemingly impossible.

What about the second problem: that propositions like 'the coin lands heads' are not equivalent to a set of points in phase space, because such propositions are compatible with different laws and the like? Again, the defender of the modal volume theory can advocate a stance of 'warranted neglect'. Because of the success of theories like statistical mechanics, it is plausible that the sorts of predictions made by statistical mechanics are among the *best* things to believe. So when I wonder whether the coin will land heads, it is best to *ignore* the possibility that the physical assumptions built into phase space will turn out to be otherwise. So I can assume that the relevant region of phase space, which represents the coin landing heads in a world that conforms to the classical picture, *exhausts* the possibilities that I need to consider when working out the best thing to believe. It is true that this set of points is not a good way of representing the *meaning* of the proposition, 'the coin lands heads', but it is a good way of determining the best degree of belief for that proposition.

Finally, the problem of vagueness is less easily dismissed, and warrants a much longer discussion than I can give here. One simple thought is that, to the extent that vagueness infects the volumes of propositions in phase space, it will infect the chances. But that might be acceptable – perhaps chances can be vague or indeterminate.

Another thought might be that for some sorts of system, vagueness in the original specification will 'come out in the wash'. The thought here is that there are lots of different precise things a vague proposition could mean, but that for any given way we make the meaning precise, it will still turn out that the chance takes the same value. This is a plausible thought about the coin. It doesn't really matter whether we count those atoms out near the edge of the coin as part of the coin or not, because it won't make any difference to the *ratio* of the volumes. The sets will change a little, depending upon whether or not you include those atoms on the periphery, but the ratio of heads-and-tossed-possibilities to tossed-possibilities will remain constant. This is the sort of approach recommended by a *supervaluationist* analysis of vague terms (see box).

Vagueness and supervaluation

The supervaluationist understands vagueness as a case where there are many precise candidate meanings for a term, and no way to settle which is the correct one. But a sentence using such a term can still be true, if it is true on all the candidate precise meanings. It can be false if it is false on all the candidate precise meanings. And it is indeterminate if it is true on some, false on others.

So by *any* candidate meaning of 'bald', 'Yul Brynner is bald' comes out true. So the sentence is true, *simpliciter*. By *any* candidate meaning of the word, 'Chewbacca is bald' comes out false. So the sentence is false, *simpliciter*. But there may be borderline cases, such as 'André Agassi was bald during *t*', where *t* is some early stage of his career. This could be true on some candidate precise meanings, and false on others. So the sentence is indeterminate.

Clever though this technique may appear, it brings with it significant further costs which are beyond my remit to discuss here. See Sorensen (2008) for further discussion and reading suggestions.

25 Chances and determinism

Recall that in §4 I raised the possibility of an objector to my proposed characterisation of chance. The objector took the view that chances are objective, and hence should have nothing to do with what evidence is available. Moreover, as a result, the objector took chance and deterministic laws to be incompatible. If there are deterministic laws, then the future is not 'open' in the way that the existence of chances seems to imply.

How would the objector think about chance in statistical mechanics? The objector cannot understand chance in terms of the modal volume theory, precisely because of the problem related to measure zero events. So understood, the objector's 'chance' that the coin lands heads would be something like:

$$\frac{\text{volume}(\text{Heads} \cap S)}{\text{volume}(S)}$$

where S is the region of phase space that represents the current state of the world when the coin is tossed. Crucially, however, the objector understands S to be the current *micro-condition*. As we have already observed, micro-conditions have measure zero. So if the objector attempted to use the modal volume theory, they would obtain the unwanted result that the chance is undefined.

Rather, the objector would need to appeal to some other way of understanding chance. Intuitively, the thought would be that S is either a world that is determined to lead to an outcome of heads or not. So if S is a subset of Heads, the chance is one. If not, the chance is zero. (Keep in mind that S has only one member, so S cannot partially overlap Heads.)

But what would the objector have to say about the understanding of chance that the modal volume theorist favours: one whereby S refers to the *macro-condition* in which the coin is tossed? The objector would presumably reply along these lines:

If one understands classical statistical mechanical (CSM) chances as arising from a probability distribution over a coarse-graining [into macrostates] of initial conditions ... then it seems as if such chances are just expressing our ignorance over the actual, precise initial conditions. For instance, if the gas molecules in a (sealed, elastic-walled) box are in fact at precise distribution d, and if the Newtonian laws entail that d evolves into d', then this fixes what will be and identifies the real causal processes involved. The CSM chance is only telling us how to guess at what will be, and how to guess at the sort of causal process involved, absent the precise knowledge. It is merely a statistical average.   (Schaffer 2007: 119)

As I indicated earlier, I cannot fault this understanding of chance as incoherent. Rather, I find fault with the concept as one worthy of philosophical inquiry. The sort of chance that interests me is the sort that might actually be used in a scientific theory. That is not to say that it is easy to learn the chances – they may be forever beyond our ken. But the sort of evidence that chances are based upon must be *available* evidence, in some sense.

There are good reasons to think the micro-condition of the world is unavailable to us, in all plausible senses. To know the precise micro-condition of the world, we would need to overcome two deep and abiding limitations on our epistemic abilities: we would need to have an infinite memory capacity to store all the information required, and we would need to have perceptual capacities that were not susceptible to noise and distortion – so as to enable us to make perfect measurements. Given these limitations, it is not conceivable that agents like us could know the world's complete micro-condition. (As things stand, it is extremely difficult for us to know the macro-condition of the world – but it is at least conceivable.)

But suppose that we relax our understanding of 'availability' to a very great extreme. Then we can still embrace Schaffer's understanding of chance as one limiting case of the more general concept. Schaffer's chance is what you get if you allow the 'available' evidence to include evidence that we concede no human will ever actually have, but which it is perhaps *conceivable* that an epistemic agent *could* have: someone like our hypothetical deity, for instance. At this limit of epistemic powers, indeed, chance and determinism are incompatible. But that is just an upshot of a particularly extreme conception of chance.

Is the disagreement between Schaffer and myself merely terminological? Although it begins that way, Schaffer's conception of chance entails that statistical mechanical probabilities are not chances, and hence that they are ill suited to feature in physical explanations. Hence, Schaffer claims that statistical mechanics gives *probabilities of explanation* rather than true probabilistic explanations (2007: 119). So Schaffer's extremist conception of chance fosters an extremist understanding of physical explanation. And in turn, I suspect this will further affect Schaffer's understanding of counterfactuals and causation. So the debate is not merely terminological, because of the connections between chance and related concepts.

Schaffer's characterisation of the chance role

Schaffer (2007) gives a number of constraints that our concept of chance should meet, and argues that all species of chance that are alleged to be compatible with determinism utterly fail to meet his constraints. Unsurprisingly, perhaps, I think he has the wrong constraints, as follows.

First, he proposes the Principal Principle as a constraint. I accept this, while disagreeing with Schaffer as to how to characterise admissible evidence. See §4 where I have already discussed the relationship between Lewis's Principal Principle and my approach to chance.

Second, Schaffer suggests that non-trivial chances should concern only future events. I simply reject this claim – it is only plausible given an understanding of chances that is derived from a popular, though scientifically unwarranted, ontological picture. There are plenty of contexts in which we use chance talk about the past, and that talk seems to be perfectly continuous with our talk about chance in the future. I return to the issue of the relationship between chance and time in Chapter 11.

Third, Schaffer suggests two principles relating chances to possibilities. Focusing here on the more demanding of these, his 'Realisation Principle' states:

If chance$\langle p_e, w, t\rangle > 0$, then there exists a world w_{ground} such that: (i) p_e is true at w_{ground}, (ii) w_{ground} matches w in occurrent history up to t, (iii) w_{ground} matches w in laws.   (2007: 124)

'chance$\langle p_e, w, t\rangle$' is the chance that a proposition p_e, stating that event e occurs, is true in a world w, at a time t.

This principle seems very plausible, but clause (ii) is not quite right. First, by being framed in terms of matching in *history*, it builds on the assumption rejected above that non-trivial chances always concern the future. More seriously, it supposes that w_{ground} matches w in *all* occurrent history. But all occurrent history will vastly outrun the available evidence. I suggest that clause (ii) should be amended to read 'w_{ground} matches w in available evidence up to t'. So understood, provided the micro-condition of the world is unavailable – as it surely is, except in the most fantastical of hypotheticals – then there is no conflict between this constraint and the existence of deterministic chances.

Fourth, Schaffer proposes an Intrinsicness Requirement. The thought here is that, in intrinsically similar chance-trials, the chances should be the same.

If e' is an intrinsic duplicate of e, and the mereological sum of the events at t' is an intrinsic duplicate of the mereological sum of the events at t, then chance$\langle p_e, w, t\rangle$ = chance$\langle p_{e'}, w, t'\rangle$.   (2007: 125)

This certainly seems appropriate in many scientific contexts where we are talking about the chance of a particular experiment turning out one way or another. It also appears to be in tension with my account of chance, because trials which are intrinsically similar might have different evidence available about their likely outcome. Suppose that I am interested in investigating the half-life of a particular isotope. I have performed this experiment many times before. But before commencing this trial, I receive news from a time traveller from the future that this trial will proceed uncommonly fast. Does that change the chance of decay? In the most straightforward sense, yes: this trial is more likely to exhibit decay in a given period of time than others that are intrinsically just like it.

How do I account, then, for the plausible view that what I am interested in investigating in this trial is something that is the *same* in this trial as in all duplicate trials? It seems possible that it is because we are sometimes not interested in the single-case chance, but are interested in investigating the chance associated with a *type* of trial. What is the available evidence with respect to all trials that are of this *type*? It is the *same* for all intrinsically similar trials. So that chance – the chance associated with this intrinsic type of trial – is not changed by the peculiar evidence associated with this case. In support of this thought, it seems coherent to imagine myself thinking: 'Well, with regard to this trial, of course the chance is high that it will decay very quickly. So if I were gambling on this decay time, I'd bet on a quick decay. But as a scientist my job is simply to treat this trial as one data point among many, for the purposes of working out the chances for trials of this type. So in that sense, this trial has the same chance as any other.' (Also, as I suggested earlier, the sort of evidence that is admissible with regard to a chance hypothesis may vary with context; this sort of example could no doubt be used to illustrate such context-sensitivity.)

Fifth, Schaffer claims a link between the laws and chances, of the following form:

Lawful Magnitude Principle: If chance$\langle p, w, t \rangle = x$, then the laws of w entail a history-to-chance conditional of the form: if the occurrent history of w through t is H, then chance$\langle p, w, t \rangle = x$.   (2007: 126)

Again, Schaffer's framing of this constraint builds on an assumption that chances are future-directed. Better, for my purposes, would be if the laws entail a 'present-to-chance' conditional. But putting that dispute aside, the interesting issue is whether or not we can have two worlds that agree in their laws, but which have different history-to-chance conditionals true in them.

If there are deterministic laws, then it is true that there are propositions that will be true of the world that are conditionals of the form 'If the complete state of the world at *t* is *H*, then future event *e* will/will not happen.' That is, the laws will be enough to entail deterministic conditionals of this sort. But it will also be the case that, given a specification of the epistemic circumstances of an agent, there will be a physical limit to what evidence is available to that agent. So there is no conflict between supposing that the present state of the world entails what evidence is available in a given agent's circumstances, and hence further entails what are the chances.

Finally, Schaffer claims that if chances are causally relevant, they must occur in the temporal interval between cause and effect:

Causal Transition Constraint: If chance$\langle p_e, w, t \rangle$ plays a role in the causal relation between *c* and *d*, then $t_e \in [t_c, t_d]$. (2007: 126)

This is plausible enough, and I'm happy to accept it as a constraint. But this has no bite until we further stipulate that certain chances *are* causes. This may be problematic, and I discuss this issue in the main text, §26.

26 Sceptical responses

The modal volume account works relatively well at *explicating* the concept of chance. Chances do seem to be facts about possibilities. But if this is to be what we want it to be – a substantial theory of what *makes* a matter of chance – we need to ask for more than just explication. Is it plausible that facts about possibilities, and about a measure over possibilities, make it the case that the best thing to believe is to match one's degree of belief with the ratio of volumes, along the lines of M_1?

There are some powerful reasons to be concerned that this account, as a theory of chance-makers, gets things badly wrong.

1. First, as we've already seen in the discussion of chance and determinism in the previous section, some people will be concerned by the modal volume theorist's *attitude* towards phase space. Phase space, remember, is a representation of mere *possibilities*. In particular, when we use phase space to represent a macro-condition of the world, we are using it to represent

various possibilities that we are unable to distinguish between, given our current state of knowledge. The blob in phase space that I described contains a single point that perfectly represents the actual world, but it also contains infinitely many points that represent alternative possibilities. Those alternative possibilities are in some sense *not real*. They are merely ways I might think the world to be; they are not ways anything *actually is*.

Given that these representations are in the blob because we are unable to rule them out, the blob appears to be a representation of our ignorance. It contains a representation of the truth, but it also contains all manner of things that are pure fictions. The bigger the blob, the greater our ignorance. How can it be that *measuring* our ignorance can yield an objective truth about the probability of a process occurring? (Note that these concerns would also apply to Chance-4 if we, contrary to my recommendation, took it to be an analysis of chance.) In a related spirit, someone might turn this into a concern about the causal efficacy of chance. In reality, either the air will rush in or it won't (they might say). And whatever it does, it is *determined* to do so by the mechanical laws. Therefore the 'chance' of which we speak when we claim that the air is probably going to leak out, rather than in, is merely an expression of our degree of belief in the proposition that it is going to leak out. This 'chance' cannot be a causal explanation of why the air leaks out – our *beliefs* cannot be causing the air to come out!

2. A second objection to this approach is that it fails to *explain* the crucial normative role of chance. Look again at M_1. It is intended not merely as a way of calculating the chance, but it purports to be a claim about what *makes* a matter of chance. Using my gloss on chance as the best thing to believe, given the available evidence, we can re-write the force of M_1 as follows:

If the best degree of belief that p, given q, is x, what makes this the case is that, using the standard measure of volume on phase space:

$$\frac{\text{volume}(p \cap q)}{\text{volume}(q)} = x.$$

But what does the standard measure of volume have to do with what is best to believe? If we tell some ancillary story, about the inductive success we have had in using this measure, then we will seemingly have changed our tune: what makes something the best thing to believe is its inductive success, not its being defined by the formula above.

Can something be said about the measure of volume which transparently links it to epistemic success? If we could do that, then we would be able to show why facts about the measure constitute facts about what is best to believe. But without that, this proposal seems like a brute stipulation, rather than an enlightening explanation.

The idea behind this objection is not that we should disbelieve the results of classical statistical mechanics. Rather, it is that we have misunderstood the *relationship* between chance and the conceptual apparatus of statistical mechanics. Chance is something like the best thing to believe. Calculations about volumes of phase space may be very helpful in working out the best thing to believe. But *those calculations are not themselves the chances.* How could they be?[3]

One way to try to make the modal volume theory more plausible, in light of the above criticisms, would be to examine how we learn about chances, and to try to show how facts about measures over possibilities might play a role in our learning process. That is the approach I will take in the next section.

27 How do we initially grasp the measure over possibilities?

The modal volume theory is a descendant of the classical account of probability, associated most strongly with Pierre-Simon Laplace. The classical account says that a conditional probability that p, given q, is a *ratio* of possibilities. Here is how the idea was thought to work out in practice: what is the probability that a die lands 6 when tossed? It will be the ratio $\frac{1}{6}$, because

3 In my presentation of this objection, I am influenced by a number of critics of related theories. For instance, Eagle (2004) writes:

> We wish to find an analysis of probability that makes the scientific use an explication of the pre-scientific use; but this project should not be mistaken for the project of discovering a scientific concept of probability. The second task had been performed exactly when we identified scientific probabilities with normed additive measures over the event spaces of scientific theories. But to make this formal structure conceptually adequate we need to give an analysis of both the explicandum and the explicatum.
>
> An analogy might help here: as standardly interpreted by the Kripke semantics, the $\ulcorner \Box \urcorner$ of S5 is a precisification of the pre-theoretical concept of necessity. However, merely giving various conditions on the box that makes it behave in roughly similar ways to necessity does not yield an analysis. The role of possible worlds in the Kripke models itself cries out for philosophical attention, in just the same way as the original pre-theoretical concept of necessity did. (2004: 372)

The modal volume theory does go somewhat beyond simply identifying a scientific concept of probability as an interpretation. But it does not go *much* beyond this. And as I have tried to show, it is inadequate to that task.

See also David Lewis's complaint against non-actualist accounts of chance:

> Be my guest – posit all the primitive unHumean whatnots you like ... But play fair in naming your whatnots. Don't call any alleged feature of reality 'chance' unless you've already shown that you have something, knowledge of which could constrain rational credence. (Lewis 1994: 239–40)

My concern here is very close to Lewis's: that the modal volume theory does not show us how knowledge of modal volumes would constrain rational credence.

(i) there are six possible ways for the die to land, and (ii) of those six ways it could land, only one is a way in which it will land 6.

Though the answer given is very plausible, as a general theory of chance, the classical account has very shaky foundations. Why, in particular, should we accept that there are precisely six possible ways for a die to land? Admittedly, one *can* divide the possibilities in that way – ignoring subtle differences between different ways the die can land, and paying attention only to which face is up. But what makes that the *correct* way to divide up possibilities? Why not say there are two possibilities: (1) it lands six or (2) it doesn't? If that were a legitimate division, we would get the result that the chance of a six is ½!

More realistically, given the continuum of spatiotemporal trajectories the die could take, there are *infinitely* many possible ways that a die could land. A theory that defines probabilities by ratios of possibilities is going to be incapable of handling infinite sets of possibilities, and thus is going to be useless for the real world.

Classical theorists fiddled with the theory in various ways to try to accommodate the unruly expanse of probabilities. One persistently misguided feature of their approach was that they assumed that it was possible to work out the relevant classes of possibilities a priori. But that makes the calculation of a chance into a purely mathematical activity: something that can be derived without experiment. This was a mistake. Working out the chances requires empirical investigation.[4]

So the modal volume theory has two key innovations over the classical theory: first, by using a measure over possibilities, the modal volume theorist has the right mathematical apparatus to tame the infinity of possibilities. Second, the modal volume theorist stresses that determining the correct measure is an empirical activity. When we do experiments to determine the half-life of an isotope, we are refining our understanding of that measure. When we toss a coin to determine whether it is fair, we are testing a measure, with the possibility that we will discard it for an alternative.

If we take this story seriously, however, we need some sense of how we could manage to learn about the correct measure over possibilities. This is at least prima facie mysterious. We only ever experience what happens in *one* world, so how can we learn anything about the relative volumes of different sets of possibilities?[5]

One thought that can start to alleviate this concern is to try to show that we have some rough grasp on the correct measure over possibilities,

4 See Hájek (2010: §3.1) for a brief and instructive discussion of the classical theory.
5 For an influential presentation of this problem, see van Fraassen (1989: 84–5).

which comes prior to our investigations of frequencies and the like. Perhaps given this sort of primitive foothold, we can leverage our way to a full-fledged understanding of probabilities through further interaction with the world.

Consider an analogy with our knowledge of physical space: physicists have good reason to think that a particular mathematical measure is the correct way to describe spatial volume. This measure is implicit in the mathematics of physical theories like general relativity. In our ordinary practice, we have no grasp of the mathematics involved in assigning volumes to regions of physical space, and this might make it seem mysterious how physicists ever learned about the correct measure. But still, we have a good basic grasp on how to estimate volumes, and see that one region has a greater volume than another. For instance, this is relatively easy when the regions are of similar shape, convex, and you can see that one region contains points that are very far apart, and the other does not.

Likewise, imagine in turn the sets of possibilities in which each of two propositions are true. Sometimes it seems relatively easy to estimate relative probabilities, and see that one proposition is more probable than another. It is particularly easy when one set entirely *contains* another. So, for example, the probability that there is a dog in my kitchen is surely greater than the probability that there is a Dalmatian in the kitchen. And this is because, whatever volume the latter proposition occupies in phase space, the former proposition must overlap it entirely, and it must include some other possibilities as well. So it must have a somewhat larger volume.[6]

This sort of judgment might be reasonable, even if we remain very uncertain about what the particular explanatorily powerful measure will turn out to be. With this sort of confidence about how the measure will turn out, we seem to be getting a basic grasp of what the measure is.

6 Actually, empirical psychology has shown that even estimations of probability in very easy cases like this can be prone to serious error. For instance, in celebrated work in cognitive psychology, Daniel Kahneman and Amos Tversky have shown that even subjects who have taken advanced graduate study in statistics and probability will sometimes rate a conjunction of propositions as more likely than either conjunct.

In one study, Tversky and Kahneman (1983) presented subjects with the following text:
Linda is 31 years old, single, outspoken and very bright. She majored in philosophy. As a student, she was deeply concerned with issues of discrimination and social justice and also participated in anti-nuclear demonstrations.
Subjects were asked to rate the probability of various propositions about Linda. Approximately 85 per cent of respondents rated it as more likely that Linda is a bank teller and a feminist than that she is simply a bank teller. But any such assignment of probabilities violates the axioms of probability theory. This result was relatively robust in both naive and sophisticated respondents.

So allowing that we might have a rough, initial method of grasping the idea of the volume a proposition takes up in modal space, the hard question remains: even if we can identify some sort of measure in the right sort of state space, how do we learn whether or not it is the correct measure of chance? How do we learn, in particular, that it is the sort of thing which should determine our degree of belief?

Here is the best sort of rationale that I think can be given.

Recall the deity, making a detailed memento of the world. Our own activity of inquiring into the world is, in a limited way, analogous to this. Because of our epistemic limitations, we are unable to identify which *particular* world we are in. Instead, supposing inquiry goes as well as is possible for us, we can identify a set of epistemic possibilities in which we take ourselves to be located. Given the available evidence, there are still a multitude of ways the world might be.

Each of those epistemic possibilities is populated by an individual who is one of your 'epistemic peers'. They each have the same available evidence as you do and the same reasoning powers.

Imagine that you are about to toss three coins in the air simultaneously. Some of your epistemic peers live in possible worlds where the result will be three heads. Some will get three tails, and some will get results in between. The problem is, among these epistemic possibilities, *you do not know which is you.* Because of this ignorance, when you try to form a rational belief about what will happen in future, you are attempting to adopt an epistemic policy that is rational for yourself *as well as* for your epistemic peers.

Suppose, first, you adopted the belief that, with probability 1, three heads will come up. We know that if all your peers adopt that belief, many of them will get things wrong. There are – intuitively – a 'good number' of possibilities in which three heads is not the outcome. So this does not seem like a rational policy. 'Many' of your epistemic peers will do badly on this strategy.

So suppose instead that you form a belief that is informed by a more sensible measure on state space. On this measure, given that three coins are tossed, the chance of three heads is 0.125. Accordingly, you believe this to be unlikely. I then invite you to bet, at even odds, on three heads. That is, for a stake of \$1, you will win \$1 if three heads come up, and you will forfeit your stake otherwise. Assuming you are motivated to win, and have no particular love of risk, you would decline a bet like this. The prize is too small, given the low probability of success.

Not all of your epistemic peers will derive an optimal outcome on this strategy. All of you will forgo the bet, but some of you *would have won.*

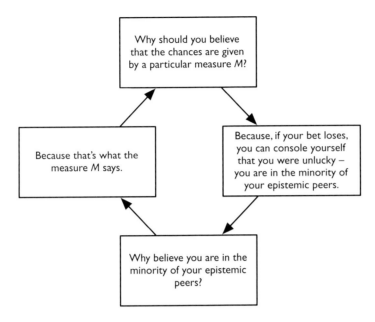

Figure 6.3 The circular chain of justification involved in possibilism.

Each epistemic peer who declines the bet, where the bet would have won, has been failed by the epistemic strategy. So can we say something about the relative degree to which the strategy has succeeded or failed? Could we, for instance, compare the *number* of successful to unsuccessful epistemic peers? Unfortunately not. There are *infinitely* many who will succeed and *infinitely* many who will fail.

There is, however, some sense in which 'more' of your peers will succeed than fail. That is, using the postulated *measure* over possibilities, we will find that the set of epistemic peers who are unsuccessful has measure 0.125, relative to the total set of epistemic peers. So if you decline the bet and find that three heads indeed come up, you will be able to console yourself that you have been *unlucky*. In some sense, 'most' of your epistemic peers have succeeded by declining the bet. You just happened to be in the minority.

As a justification for believing that a given measure determines the chances, however, this is manifestly circular. The reasoning is illustrated in Figure 6.3.

We will need to do better than this if we are to show that the modal volume theory is a correct account of what makes a matter of chance.

28 How do we make better estimates of chances?

Perhaps the modal volume theory can be vindicated by showing how we can be motivated to *change* our beliefs about chance. Changing our beliefs involves changing the measure we use over state space. If we can show why these changes would tend to lead us towards greater epistemic success, then we can be reassured that the modal volume theory really does guide us to the best thing to believe.

Suppose that you were in the 'unlucky' minority of epistemic peers, and the three coins do come up heads. You would probably not be rationally required to revise any beliefs at this point. Rather, you would simply shrug your shoulders and put it down to bad luck.

Now suppose that you repeat the experiment many hundreds of times. Surprisingly often, you get three heads. It seems that these coins produce the outcome three heads approximately 25 per cent of the time.

At some point, it will become reasonable for you to revise your belief. Why? Presumably because you should not assume that you are extremely unlucky. You should assume that you are fairly typical, and you simply have got the wrong way of estimating the chances.

So you change your mind about what the measure is. Let's say you work out some new way of apportioning beliefs such that the chance of getting three heads is indeed close to 0.25.

There are two possible back stories to this change in belief.

Unlucky If you are *really* unlucky, you are wrong to revise your beliefs in this way. The true chance for three heads is – let's suppose – 0.125. The only consolation is that *very very few* of your epistemic peers will be deceived in this way. Because it is very unlikely that you will observe a frequency of 25 per cent over hundreds of trials, only a very few of your epistemic peers are going to be exposed to this misleading evidence.

Not unlucky It could be that the chance of three heads really is around 0.25. In that case, your initial theory of chance was incorrect, and this has been revealed by the trials. Moreover, it will be revealed to the *vast majority* of your epistemic peers. Your change of belief is a change that will occur to most of them, and it will be rational for all of those peers.

So the lesson here seems to be: if you revise your beliefs about the measure on the basis of an assumption that you are not especially unlucky among your epistemic peers, then *provided that assumption is correct*, you will be led to revise your beliefs in ways that better conform to the chances.

But again, in trying to explain how it is that we will cotton on to the correct measure, we have made an appeal to a chance concept that itself needs to be cashed out in terms of the correct measure. It is less manifestly circular in this case, but it still looks suspicious:

- Why believe that I should change my beliefs about the chances, reflected by dropping measure M, and adopting measure $M^\star$?
- Because, if you retain M, you need to believe that you are very unlucky. $M^\star$ is a measure that better fits with the assumption that you are not very unlucky.
- Will this change in beliefs lead me to the truth?
- It will, if you are right in your assumption that you are not unlucky.
- What does it mean to be not unlucky?
- It means your epistemic peers have a large measure, on the preferred measure.

So what makes a measure preferred, on this story, is its fit with your evidence. If a measure makes your evidence extremely unlikely, then you ought not to believe it. But this is a serious problem if the measure is supposed to be an account of chance. Chance is what you ought to believe, given the available evidence. So for the very unlucky person, who is in a world where the evidence is misleading, he will come to believe measure $M^\star$ which, *ex hypothesi*, is the *wrong* measure.

A defender of the modal volume theory could respond by saying:

Yes, what you have argued is correct. But this just reveals the inadequacy of your characterisation of chance. Right from the get-go, you tried to define chance as the best thing to believe. But that conflates *epistemic* probabilities with *real* chance. Your objection merely shows that real chances come apart from what it is best to believe.

But this response makes the modal volume theory altogether unbeliev-able. Without a conceptual tie between chance and our epistemic practice, what role does chance play? What reality does this measure have, that gives it a 'grip' on what happens in the world, such that I should have any beliefs about it? Not only does chance lack the relation to optimal belief that I have posited, it allows for radical, persistent, and widespread divergence between the chances and what it is rational to believe. In that case, why should we expect this measure to feature in our physical theories at all?

The modal volume theory, as a metaphysical account of chance, must be rejected. I now turn to some alternative approaches to chance, to see if they can do any better.

Propensities, and other possibilist accounts

I have not here discussed explicitly other possibilist accounts, such as Popper's propensity account of chance (Popper 1957). Although propensity theorists may not explicitly mention a measure over possibilities, it is clear that their accounts are intended to have some degree of modal force which strongly suggests that they are susceptible to the same arguments discussed here. Generalising somewhat, the distinctive features of a propensity account, as compared to the sorts of possibilist features mentioned here, are that propensities have additional connotations, over and above the idea that probability is something like a degree of possibility. For instance, most propensity accounts aim to vindicate the idea that chances are features that supervene on the intrinsic properties of a chance set-up. Propensity accounts might also hope to capture the idea that chances are causally efficacious. While there are some chances for which these ideas are plausible, it proves to be a handicap to build them into the concept from the outset, so by considering weaker ideas, I show the generality and strength of the case against possibilism.

For a more focused and sustained attack on propensity accounts, see Eagle (2004).

29 Weaker versions of possibilism

Part of what makes it difficult to justify possibilism is that there are so many possible measures over the relevant space of possibilities. It is hard to see what could make one measure uniquely preferred. And even if one measure is uniquely preferred, it is hard to see how we could ever learn which measure has this special status. One alternative approach to thinking about chance employs a broadly possibilist framework, but without assuming a uniquely preferred measure. The key idea is that, *for systems with the right kind of dynamics*, we can be very confident of a particular probability distribution for the outcomes, largely independently of how we distribute our credence over the initial conditions.

To illustrate, suppose you are told that there are currently one million coins lying on the surface of the moon. You have no idea, moreover, how the coins have got there and thus how they are arranged. It could be that they are all lying with heads facing up. It could be that they are laid out in an idiosyncratic pattern such that every fourth coin is tails, and the rest are showing heads. And there are of course many other possibilities in between. Given this information, it seems implausible that there is a uniquely preferred measure over the space of possibilities which will tell you the probability of any one arrangement or the other. Imagine that a number of us have very strong – even dogmatic – attitudes about the initial conditions of the coins. We disagree with each other, and assign very different probabilities to the different arrangements, even though we all share the same relatively impoverished evidence.

However, if we are told that an asteroid has just struck the surface of the moon, causing all the coins to jump off the surface at least 3 feet each, and then to fall back down, we will have a different attitude. Now we will all be very confident that about half the coins will be showing heads, and half will be showing tails. Despite having started out with wildly divergent views about the likely initial conditions, we are all able to agree that, now, this is the most likely distribution. This suggests that there is something important about the dynamics of how coins fall and land that is important in these considerations, and that this is playing a larger role than is acknowledged by the possibilist accounts which place such heavy emphasis on the measure over initial conditions.

This idea has been developed more rigorously and systematically by Michael Strevens, who calls it the microconstancy account of probabilities. Similar ideas have been explored recently by Tim Maudlin and Aidan Lyon.[7]

The microconstancy account works as follows. First, we will stipulate that the physical laws are entirely deterministic. Second, consider all the possible initial conditions for a system containing an asteroid, a moon, and all the coins lying on the moon. Some of the initial conditions will differ in

---

7 Strevens gives a rigorous, but technically demanding treatment of the physical and mathematical details, while remaining relatively agnostic on the metaphysical issues (Strevens 2003: chap. 2). Indeed, he claims that the account is compatible with any metaphysics of probability (§1.33). In a later paper (Strevens 2011), he explicitly attempts to work out a metaphysically complete – and implicitly *possibilist* – account of microconstant probabilities as a species of physical probability.

Tim Maudlin discusses a very similar approach, but expresses doubt that it can constitute a proper instance of single-case chance, in Maudlin (2007b: §3). The above example of the coins lying on the moon is inspired by Maudlin's discussion.

Aidan Lyon (2010) argues that the sorts of probability discussed by Maudlin and Strevens must be a third species of probability – neither credence nor chance – though Lyon's conclusion that these probabilities are not chances depends upon accepting some metaphysical tenets about chance – such as the incompatibility of chance and determinism – which I am inclined to reject. So while I agree that these probabilities are not credences, I cannot agree – on Lyon's grounds – that they are not chances.

For recent criticism of these ideas, see Frigg (2011).

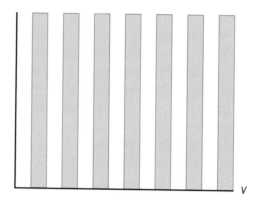

Figure 6.4 How the outcome for a single coin depends upon the initial conditions of the moon–asteroid system. The quantity *V* on the horizontal axis is the velocity with which the meteorite strikes. The shaded regions represent initial conditions which result in landing heads.

the velocity of the asteroid. Some will differ in where exactly the asteroid is coming from. Some will differ in how many coins are showing heads or tails. Suppose that one of the people interested in predicting the distribution of heads after the asteroid has struck – call her Deirdre – is nearly certain that all the coins are currently showing tails. So she gives minimal credence to the possibility that the coins are distributed any other way. Still, Deirdre is not sure of the exact direction and velocity of the asteroid. So she has to distribute her degrees of confidence over the many possible directions and velocities.

Given what we know about the physics, for each small change in the velocity or direction of the meteorite, what sort of difference can we expect in the outcome, for any given coin? The *actual* physics is too hard to do here, but the following is a workable simplification. We have good grounds for thinking that there will be the following sort of relationship: if the meteorite lands a bit harder, the coin will fly up a bit more, completing a greater degree of rotation before it lands. So if the velocity goes up by a very small amount, it will make no difference. If it goes up by enough, the coin will complete another half-rotation, and will land on a different side. If it goes up yet more, the coin will complete a full rotation, and will land on the same side. Figure 6.4 illustrates the way the outcome depends upon these small variations in velocity.

Notice that there is a roughly constant ratio between the proportion of initial conditions that lead to heads versus those that lead to tails. For any reasonably large continuous set of initial conditions you consider, you can expect about half of those conditions to lead to heads. This is, roughly, what Strevens means by 'microconstancy'.

Of course, things are more complicated than this: landing at some angles will lead the coin to bounce much more than landing on other angles. So if the coin completes only a tiny fraction of a turn, it might still be enough to lead to it landing on the other side, because of the effect of bouncing. We can ignore this, however. Roughly, the reason is that for any value that the other variables take, we still expect to see the same *pattern* of dependence upon the velocity of the asteroid at impact. That is, we still expect that, for any reasonably large set of possible initial velocity conditions, about half of those conditions will lead to heads and half will lead to tails. (Strevens gives a much more careful articulation of the assumptions required to make this reasoning reliable.)

Suppose that the illustration in Figure 6.4 is indeed a good representation of the possibilities for a coin that is currently showing tails. How would it differ for a coin that is currently showing heads? It would look much the same, but 'shifted along' to one side. All the shaded regions would be empty, and the unshaded regions in between would now be shaded. So the disagreement between Deirdre and the others, who have different opinions about how the coins are initially distributed, is – at least by implication – a disagreement about which of these graphs correctly represent the way the outcome depends upon the velocity of the meteorite. However, despite this disagreement, we agree with Deirdre on what the likely outcome is, after the meteorite strikes. How is this so?

It comes about because our degrees of belief about the velocity of the meteorite have an important similarity. All of us have relatively 'smooth' credence distributions over the various possibilities. That is, we do *not* tend to think that the meteorite is likely to be travelling at speeds up to 35 metres per second, but is *not at all* likely to be travelling at speeds greater than that. In Figure 6.5 two distributions of credence are shown as solid lines over the possible initial velocities of the meteor. Although these are very different distributions of belief, both will have approximately half the area underneath them shaded, and half not. That is: agents who hold these degrees of belief, and who know that the physics is as represented here, will have excellent reasons to think that the probability of the coin landing heads is 0.5.

This conclusion is robust, even if the agent thinks the coin has started out showing heads, rather than tails. Simply shifting the shaded regions across will not affect the microconstancy of the dynamics. So it will still be the case that half of the region underneath the curve is shaded. Moreover, for almost

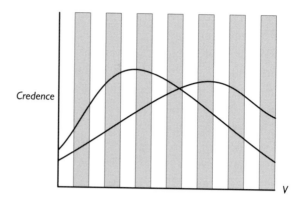

Figure 6.5 Each of the two curves superimposed on the graph for the way the outcome depends upon the velocity of the meteor represents a distribution of credence over some of the possible velocities. One agent believes the velocity is most likely to be high; the other believes the velocity is most likely to be low.

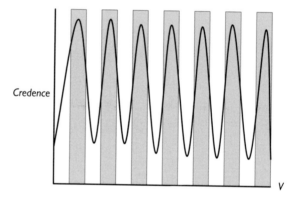

Figure 6.6 A 'jagged' credence distribution. Someone with beliefs like this about initial conditions would think that heads is extremely likely. But because the difference in velocity that leads to a different outcome for the coin is so small, it is very hard to see how any human could have evidence that makes such a jagged distribution rational.

any sufficiently 'smooth' distribution of credence over the initial velocity, the same result will hold. The only sorts of agents who won't agree with Deirdre about the likely distribution of heads will be agents with bizarrely 'jagged' distributions of credence over initial conditions, as in Figure 6.6. Because the differences between initial conditions that lead to heads and initial conditions that lead to tails are very hard for us to detect, it is difficult to envisage circumstances in which it would be rational for agents to have degrees of belief which sharply distinguish between these possibilities. So it looks like some degree of smoothness in our credence distribution is rationally required.[8]

If all the steps in the foregoing sketch are acceptable, we have managed to establish that there is a common degree of belief in heads that is *optimal*, even to agents with radically different initial opinions about how the coins are arranged prior to the collision, and even among agents with very different opinions about the most likely velocity of the meteorite. All that they need to have in common is an understanding of the way the physics of the coin depends upon the initial velocity of the meteorite and also a reasonably smooth distribution of credence over the possible velocities of the meteorite.

Problems

The account I have given above does not do justice to Strevens's account, but it is convenient for dialectical purposes because it highlights what turn out to be some genuine difficulties.

First, as I have presented it, the macroperiodicity condition – the condition of smoothness over initial conditions – is a requirement on our state of mind. That makes it seem as though these probabilities depend upon something subjective, which is quite at odds with the thought that they are physical probabilities.

One possible response is to suppose that there is some rational requirement on all agents that they adopt suitably macroperiodic distributions in cases of ignorance. In effect, this would be something like a principle of indifference.[9] So although macroperiodicity continues to be understood as a property of our epistemic states, it is now understood as compulsory, and hence loses its air of subjectivity. But this understanding of macroperiodicity raises the same concerns that were levelled at a possibilist account of chance at the beginning of §26. That is, this understanding seems to conjure chances out of our ignorance. It is hard to reconcile the thought that chances of this sort can play a role in physical explanations, if they depend for their

---

8 Strevens calls this smoothness condition 'macroperiodicity', and gives a rigorous definition of both macroperiodicity and microconstancy in Strevens (2003: §2.23).

9 It should be noted that a recent trend in epistemology has been to revive ideas like the principle of indifference (White 2010). This would seemingly support the idea that some initial distributions of credence – those that violate macroperiodicity – are irrational. But in White's recent defence of this principle, he requires it to be silent on what the chances are – it is only intended to constrain our credence when we are ignorant of the chances (2010: 163). So it is not clear that it could be used, in this context, where possibilists are trying to use our rationally required initial distributions of credence as a way to bootstrap us into a fact about chance.

chanciness on our inability to work out what non-smooth distribution of credence would serve us better.

And indeed, Strevens eschews this sort of epistemic construal of the macroperiodicity condition. To state more carefully his preferred account, he claims[10] that:

The probability of system S producing event e is p if:

1. The dynamics of S is microconstant regarding e, and has strike ratio equal to p (microconstancy).
2. The initial conditions of (almost all long series of) trials on S is macroperiodically distributed in almost all nearby worlds (robust smoothness).

(The notion of 'strike ratio' in the first clause is simply the proportion of the initial conditions which lead to outcome e.)

Note that the second clause of this sufficient condition is non-epistemic. It makes no reference to what agents believe about the initial conditions of trials on S; it merely requires that, *in almost all nearby worlds*, trials are macroperiodically distributed across the initial conditions. This is a claim about the frequencies of particular initial conditions in nearby possible worlds.

The 'almost all nearby worlds' phrase is meant to express the idea that, if things had been different, then *very likely* the distribution of trials would be macroperiodic. This is essentially the same probabilistic notion as was employed above, in explicating the possibilist approach to chance: why believe that we are relatively unlikely to see three separate coins all land heads? Because in *most* worlds that are similar to this one in their initial conditions, when three coins are tossed at least one coin will land tails.

As has been argued at some length above, this notion presupposes a measure over worlds with which to make sense of the notion of relative quantities of worlds. It is hard to see what would make one measure over worlds uniquely preferred. It is also hard to see how we could *learn* what makes one such measure uniquely preferred. So the account which Strevens has given us, although it avoids the objection that arose for the simpler presentation above – that it is overly epistemic – nonetheless invites much the same objection as was ultimately levelled at a possibilist account: that it relies upon a metaphysical fact about possibilities which seems to be too

---

10 See Strevens 2011: 359. Strevens considers a possible third clause, omitted here, which is an actual frequency clause: requiring that almost all *actual* long series of trials on S are macroperiodically distributed. This would inject a greater measure of actualism into his account, and would subject it to some of the concerns discussed in Chapter 7.

far removed from the actual world to ensure the right relationship between chance and our epistemic states.

Moreover, the notion of a measure over possibilities is necessary, not only for the condition of robust smoothness, but also for the microconstancy condition. The 'strike ratio' mentioned in that first clause is a *proportion of initial conditions*. And of course, those initial conditions are themselves possible states of affairs. There are infinitely many possible states, so it is necessary to have a *measure* over the states for the idea of a proportion to make sense. So again, for Strevens's account to work, there must be such a measure and we must be acquainted with it.

In a recent paper, Jenann Ismael (2009) argues that the sorts of probabilities which Strevens is using – ones which rely on a measure over initial conditions – are indispensable to using physical theories to derive expectations. A deterministic theory, without some way of apportioning credences to initial conditions, is compatible with practically anything happening.

So it might seem that I am being too quick to dismiss Strevens's account, which is carefully worked out, which seems to match actual practice, and which only appeals to a measure over possibilities in a very carefully constrained way. If we must have such a measure, then perhaps we just have to live with some metaphysical perplexity about its nature.

While I am very sympathetic to Ismael's point, and happily grant that the microconstancy account has many virtues, I think that the upshot of these observations is that we need to broaden our search for an adequate theory of probability. Although our behaviour may reveal that we are implicitly committed to behaving *as if* there is a correct measure over probabilities, it is extremely hard to explain what would make this behaviour correct. The obvious alternative is to examine a less metaphysically rich account of chance: perhaps chances can be grounded entirely in what happens in the actual world; or perhaps we need to abandon realism about chance altogether. These are the topics of the next two chapters.

7 | Actualist theories of chance

The idea that chances are facts about real possibilities is – to many – an outlandishly metaphysical claim. It makes chance seem more like a philosopher's fantasy than anything else. One way of avoiding the idea that chances are facts about possibilities is to attempt to identify some non-chancy facts in the actual world which constitute the basis of probabilities. What I call *actualist* approaches thus maintain that chances are mind-independent and real, but hold that they are reducible to 'this-worldly' phenomena.

30 Actualist interpretations of chance

The actualist approach is associated with David Hume's famous discussion of the concept of causation. Hume observed that the concept of causation appeared to involve a necessary connection between cause and effect: given the cause, the effect *must* happen. But upon close inspection, Hume found that there was no observable correlate of such necessary connections. Exactly what conclusion Hume drew from this is still disputed, but one very influential interpretation of his thought is that he believed therefore that causation was in reality nothing more than the constant conjunction of two types of event.[1] Seeing two events of a particular type occur together in time and space, again and again, produces in us a belief that they are causally connected, and that belief has an associated feeling that the connection is necessary. But causation itself is nothing more than the constant conjunction. The necessary connection is some sort of 'projection' by us onto the world.

In broadly analogous vein, a Humean reductionist approach to chance would suggest that the connection I have noted between chances and possibilities is not itself observable. All we *see* are events happening – be they events of radioactive decay, the deposition of a government, or some air leaking out of a tyre. We do not ever observe something like a realm of possibility. Nor do we live in a realm of possibility. So if we believe that

1 Galen Strawson (1992) is the most prominent dissenting voice regarding this matter.

there are real physical probabilities, we need to find something in the actual world that can be the basis of those probabilities.

While Hume himself was drawn to these sorts of views by his central epistemological commitment – the claim that all ideas had their origin in experience – contemporary proponents of this Humean programme do not share this epistemological view. Their foundational idea is that our concepts must pick out phenomena which are, ultimately, entirely *actual* and *occurrent*. So a probability concept need not be entirely derived from experience, but *facts* about probabilities must be grounded in the actual world.

I won't try to say anything conclusive about the motivation for this Humean programme. I think that it remains deeply mysterious why Humeans are inclined to pursue it, given it has been severed from Hume's psychological and epistemological views. Moreover, the programme's most significant exponent, David Lewis, did not even espouse it in quite such a stringent form as I have suggested. Rather Lewis wanted to maintain that there are no *purely philosophical arguments against* a Humean account of causation and similar phenomena. He did not claim that his account of causation *had* to be right. His claim was the much more subtle one that, on the basis of philosophy alone, we cannot rule out the possibility that laws, causation, and the like are entirely reducible to actual phenomena. And indeed, Lewis recognised that the existence of objective probabilities posed a direct threat to his programme (though not on purely philosophical grounds), and that the use of probabilities in scientific theories such as quantum mechanics might give us reason to reject a Humean reductionist account.[2]

At his most ambitious, however, Lewis suggests that any more metaphysical account of causation or laws, for instance, cannot be correct, for it will fail to explain how these phenomena feature in our epistemic practices. If these things are brute relations between mere possibilia, how could we come to know them and speak of them?[3] These arguments, I believe, are of genuine importance, and we cannot afford to ignore them. On the other hand there are some serious problems, in turn, with actualism.

---

2 See the foreword to Lewis (1986b) for Lewis's discussion of the programme that he called 'Humean supervenience', and his observation that it is threatened by the existence of objective chances. A powerful critique of this programme – even with this modest ambition – is given by Tim Maudlin (2007a: chap. 2).

3 See Lewis (1983a: 40; 1994: 239–40) for comments in this vein.

31 Simple actualist proposals

The simplest account in the actualist vein is one that identifies probabilities with relative frequencies of events. In all world history, for instance, how many coin tosses will there be? Suppose it is one million. How many of these tosses will land heads? Suppose it is 500,003. The probability that a tossed coin will land heads, then, according to the *simple frequentist* view, is 0.500003.

This view encounters two severe problems. The first is known as the reference class problem, and the second is what I will call the fundamental problem.

The reference class problem

The reference class problem is simply this: a relative frequency is, by definition, relative to a class of events. This is known as the reference class. In the example given above, the reference class was simply the class of all coin tosses. But we might want to know the probability that *this* particular coin will land heads. In that case, perhaps the appropriate reference class is the class of all tosses of this particular coin. Suppose this particular coin is tossed ten times, and comes up heads on six of those occasions. Then the probability that this coin will land heads is – relative to the new choice of reference class – 0.6.[4]

How can this be reconciled with the idea that events have *objective* probabilities? If that were so, then there would need to be an objectively correct reference class of events from which the appropriate relative frequency can be calculated. But it is rather implausible to suppose that there is such an objectively correct reference class.[5]

One strategy to address this problem would be to try to maximise the size of the reference class – go for the largest group possible that shares relevant similarities with the event in question. But this risks giving us less information than we might want. A patient being advised of his risk of a heart attack does not merely want to know the risk that people of his age will have a heart attack in the next six months. He wants to know the risk that *he* will have a heart attack. And he is different from the other people of his age in

---

4 Alan Hájek presents the reference class problem, amid fourteen other objections to simple frequentism, in Hájek (1997).

5 It remains open that we could say that chances are objective properties of event–class pairs. But that means we lose our grip on how chances regulate beliefs. We want to have beliefs about the events, *simpliciter*.

various ways. He might be a non-smoker. He might have a high cholesterol level. He lives in a suburb with high levels of pollution. He drinks spirits but not wine. So to get a probability that is more salient to this individual man, shouldn't we consider a much more narrowly defined reference class?

The opposite strategy would be to narrow the classes as far as possible. But this strategy, also, seems unlikely to work. The smallest possible reference class contains only one event: either the man will have a heart attack in the next six months, or he won't. So the chance will be zero or one. That is not the sort of information we are after, either.

Finally, you might think there is some way of specifying 'relevantly similar' events, such that reference classes contain all and only the events that are relevantly similar. That would ensure – it is hoped – that the reference classes are neither too small nor too large. But how are we to understand 'relevance' here? What we mean by relevance is surely: *relevant to the probabilities*. But we are trying to *analyse* the objective probability in terms of frequency and reference class. So to stipulate that the reference class is that which is 'relevant' to the probability threatens to be circular.

What is needed is some way of specifying the relevant reference class without covertly appealing to facts about chances. Perhaps there is a way to do this. This would amount to something like dividing up events into classes that 'carve nature at the joints': that is, into groups which reflect objective similarities (of the relevant sort) and do not reflect irrelevant ones. Although it might be hoped that there is some way of doing this, it remains in serious doubt.[6]

The fundamental problem

The fundamental problem – put simply – is that chances cannot be reduced to frequencies. So no attempt to identify chances with certain sorts of frequencies can be a successful identification.

---

6 It is not entirely fair to raise the reference class problem as though it is a problem solely for simple frequentism. Arguably, it affects a large range of proposals about probability. In the case of the modal volume theory, for instance, the probability was determined by considering the macroscopic state of the relevant system. That amounted to considering a set of microstates: the reference class. But what is 'the system'? Does it include only matter in the immediate vicinity of the probabilistic trial? Or does it include the entire universe? By being vague about these matters, we overlooked the reference class problem.

Hájek claims that the best solution is to adopt conditional probability as primitive. Conditional probabilities always already have a relevant reference class. Unconditional probabilities – if they exist – can be determined from conditional probabilistic information. See Hájek (2003a, b).

Of course, the frequentist might complain that the premiss of this objection – the irreducibility of chances to frequencies – is question-begging. But sometimes it is hard to avoid begging the question when your opponent's view is so obviously incorrect!

To bring out the plausibility of the premiss, return to our imagined world of a million coin tosses. It is certainly conceivable that the coin is fair, yet that more than half of the tosses land heads. That is, it is conceivable that the objective chance of heads is precisely 0.5, but that the frequency is not precisely 0.5. The simple frequentist analysis of probability must deny that this is really possible.

Even more strangely, think what happens if we know the coin is going to be tossed an odd number of times. In that case, we can be absolutely certain that the coin is not fair, because it is impossible that precisely half of an odd number of tosses will land heads. Imagine you are told that a coin is going to be tossed only a finite number of times in its life. I then ask you whether the coin is fair. If you replied: 'Well, is it going to be tossed an *even* number of times?', we would have reason to suspect that you simply did not understand the question. But on the simple frequentist view, this would be an entirely sensible thing to ask.

And this absurdity only increases if we consider cases where there is only a *single* trial. In that case, probabilities must be either zero or one, and no other value is possible. Again, this seems to misconceive the nature of chance. Surely it is possible to say of a coin that will be tossed only once in its existence that it has a probability greater than zero, but less than one, of landing heads. But this would be a conceptual error, on the frequentist account.

We should acknowledge, before rejecting the simple frequentist view outright, that it is possible for us to be confused about our concepts. Examples of this abound in the history of science. We originally thought that solid substances had to wholly occupy the entire region in which they exist. We have since found that solid substances are in fact largely empty space. The crucial thing that makes a thing solid is the nature of the forces between the constituent particles. The presence or absence of empty space between the particles is largely irrelevant. So we appear to have modified our understanding of our own concept. We have *not* stuck to our earlier understanding of solidity, which would entail the surprising conclusion that tables, for instance, are not solid.[7]

---

7 Notoriously, this idea was denied by Eddington (1929). His remark that a plank of wood 'has no solidity of substance' has given rise to numerous philosophical attempts to explain why he was wrong.

Alternatively, you could interpret this as a case where we were never confused about our concept of solidity; we simply ditched the old concept and adopted a new one. Even though, hundreds of years ago, we used the same word that we use now, the actual *concept* associated with the word has changed. That seems a touch artificial, but even allowing that to be the case, there is still the question of what is the *more useful concept*, in light of our best understanding of the world. The reason the old solidity concept has gone by the wayside is that it cannot be accommodated in our most explanatory theories of the world.

So the fundamental objection, as stated, is perhaps too quick. While it is true that there is a mismatch between the simple frequentist account of chance and what we might have thought chance to be, it could be that this is a case where we are in need of conceptual reform. It could be that the simple frequentist account delivers a more useful concept than the concept that we inherited from our parent culture.

Simple frequentism, however, does not deliver a concept more useful than our pre-theoretical understanding of chance. Indeed, simple frequencies serve a quite different purpose, and appear unable to play the role occupied by chance in scientific inquiry.

What, then, is the role of chance – and what is the related role of frequencies? Crucially, frequencies are used as *evidence* for chances. We do not look upon frequencies of events as constitutive of the chances, but rather as an imperfect guide to what the underlying chances might be. A slight mismatch between our chance beliefs and the observed frequency is usually fine. A big mismatch is usually evidence that our belief about the chance is incorrect. When we toss a coin a million times, and find it comes up heads very nearly, but not precisely, half the time, that is a good sign that it is a fair coin. If it comes up heads 75 per cent of the time, that is much more worrying. Perhaps the coin is not fair after all. But note: the frequency is not *conclusive*. It is possible, even if the coin is fair, for it to come up heads much more than half the time. Indeed, it might even come up heads *every time* it is tossed. That is of course very unlikely, but it is a possibility we do not – and should not – rule out.

This problem, in various guises, arises for all actualist proposals about chance. That is to be expected, given that the actualist is trying to provide a substitute for our concept of chance – something *unlike* the naive, untutored concept. So we have to expect some differences between our prior conception and the actualist alternative. The question is not whether the actualist can ever eliminate the gap between our untutored concept and a reductive proposal – that would defeat the purpose of actualism. Rather, the question is whether the concept can be reformed in such a way that it serves the necessary purpose better. Can it make better sense of the way we use chances in scientific theorising? Can it make better sense of the way we come to learn things about chance? And can it do this without any other costs or disadvantages?

32 Sophisticated actualist proposals

More sophisticated Humean proposals, such as David Lewis's theory of chance, analyse physical probabilities as part of a package deal with the laws of nature. The theory is much less clearly at odds with our intuitions than proposals such as simple frequentism, but it nonetheless can be shown to suffer the conceptual problem. To show this, we will have to take a brief detour via Lewis's account of a law of nature.[8]

What are laws of nature? Stereotypical examples might include the grav- itational law which states that massive bodies attract each other with a force that is inversely proportional to the square of the distance between them. Or there might be chemical laws about the way metals and acids react. Or there might be laws of population genetics which state that certain patterns will always be observed in the distribution of genetic alleles in a population.[9] In general, then, laws are truths that hold very generally – perhaps universally – and which hold, in some sense, 'of necessity'. It is part of the very nature of the world, or of the things in it, that the laws are true of these things.

In contrast, there might be other true generalisations which are merely accidental. Although they are universally true, and have a similar general form to laws, we don't believe that they are true of necessity. Things could easily have been otherwise, we might think. For example, no head of state has ever declared war while riding a unicycle. But this is surely not a law of nature. It could easily have turned out to be false. Or to take another example, there has never been a sphere of gold 100 metres in diameter or larger. There almost certainly won't be any such sphere of gold that size for

---

8 Lewis introduced his account of laws in Lewis (1973b: 73–4). Subsequent refinements were made in Lewis (1983a: 41–2) and Lewis (1994: §4).

9 The reason I say 'might be' is to stress that we do not have indubitable knowledge of what the laws are. Indeed, we might never have such knowledge. Often scientists put forward claims that particular sentences are laws, but such claims can of course turn out to be false. Being a law, then, is not just a matter of being *regarded* as a law. There must be something about the world that makes it true that something is a law.

all time. Suppose there never is such a sphere. Nonetheless, it is surely not a law of nature that all spheres of gold are less than 100 metres in diameter. Contrast this with a similar statement about uranium-235. The nature of this isotope of uranium is such that, if you were to try to create a sphere so large, it would almost certainly undergo nuclear fission, and a nuclear explosion would result. So it is a lawlike truth that all spheres of uranium are less than 100 metres in diameter. Or, more precisely, the correct law would be something to the effect that any sphere of uranium-235 as large as that has only a very small probability of enduring for more than a fraction of a second.

For the actualist, laws and probabilities pose a similar problem. Both seem to point beyond what is *actual* to events that are merely *possible*. A deterministic law indicates that there are no other possible ways for things to happen. A probabilistic law – according to the naive interpretation advanced in §24 above – would entail that an event of a particular type occurs in a certain proportion of the possible ways things might happen. The Humean actualist programme, remember, is to try to substitute a purely actual phenomenon for such suspicious, possibility-involving concepts.

Lewis's suggestion, adapting an idea first put forward by J. S. Mill and F. P. Ramsey, is that laws are true generalisations that stand in a specially powerful explanatory relation to all the other contingent facts about the world. The way probabilities feature in this account is rather difficult to grasp, so we shall first describe the account assuming there are no chances.

Suppose you tried to write down all of the contingent truths. You could leave out truths of mathematics and logic, because these are strictly necessary truths, but you would still have an awfully long list – infinitely long, in fact. The list would cover truths about the locations of particles (much like our deity's memento), but it would also include truths about rabbits, humans, solar systems, typhoons, and so forth. Some of the truths would be about particular individuals, and others would be generalisations about all the individuals of a given type. There may be redundancy in this list of truths, but we are not yet concerned about that.

From this messy list, we could adopt a variety of strategies to prune it down into a more manageable body of truths. Lewis's thought is that if we pruned it down to get the set of truths which had an optimal balance of two properties, then that set would be the deterministic law statements. The two properties that we would balance for are: *simplicity* and *strength*.

The idea of simplicity is relatively familiar. One way the set of sentences can be simple is if they have simple subject matter. If the statements are about just a small number of relatively basic kinds of thing rather than about a very large number of relatively non-basic kinds of thing then they will be more simple. For instance, compare a set of sentences that is entirely about massive objects, charged objects, and forces. This is relatively simple subject matter, because the things mentioned are relatively basic, and relatively few in number. In contrast, consider a set of sentences that is entirely about: French provinces, rotten cheeses, left-hand gloves, things which are blue before the year 2000 or green thereafter, and 170 other miscellaneous entities. These sentences are more complex, because of their more diverse subject matter.

A second way for a set of sentences to be simple is in the types of relationships they ascribe to things. If the statements ascribe simple relations between things, rather than complex relations, this will count for their simplicity also. So a statement that all protons attract electrons is describing a simple (though somewhat imprecise) relationship between two types of thing. A statement that, for the entire day on the 1st of April, in the year 13 CE, all left-wing politicians were seventeen thousand metres or less distant from a third cousin of Caligula describes a rather more complex relation between left-wing politicians and Caligula.

Strength is something like the information content of a set of statements. When we try to be informative, we usually have to do so at the cost of simplicity. 'Some massive objects exist' is a wonderfully simple statement, but has little to recommend it in terms of strength. A sentence like:

To Enoch was born Irad; and Irad was the father of Mehujael, and Mehujael the father of Methushael, and Methushael the father of Lamech[10]

does rather better for information (supposing it to be true), but less well for simplicity. A sentence like:

But Orpheus, the son of Oeagrus, the harper, they sent empty away, and presented to him an apparition only of her whom he sought, but herself they would not give up, because he showed no spirit; he was only a harp-player, and did not dare like Alcestis to die for love, but was contriving how he might enter Hades alive; moreover, they afterwards caused him to suffer death at the hands of women, as the punishment of his cowardliness[11]

does better yet for strength, but yet worse for simplicity. And so on. But some statements manage to convey a lot of information while still being simple. For instance, a sentence derived from Newton's *Principia* such as 'Force = mass × acceleration' tells us something that applies to *all* massive objects.

---

[10] Genesis 4:18.   [11] Plato, *Symposium*, 179d2–e1.

Given that there are many massive objects, and given that the relationship it describes is not trivial, this is a very informative statement. Given it is simple also, it is a good contender to make it into the optimal set of statements.

Lewis claims that if we pursued to its ideal limit this strategy of seeking after the set of statements which best combined the properties of strength and simplicity, we would get all and only the laws. Laws contain a great deal of information – hence they do well for strength; and they are also usually relatively simple – they typically state general relationships rather than a long list of particular facts, for instance.

This approach is known as the 'best-system' account of laws, because it takes laws simply to be the sentences that feature in the best systematisation of all the contingent truths. What is especially notable about it is the way it avoids any special *metaphysical* story about what gives the laws their sense of necessity. The laws are – in the end – just truths like all the others. But the reason we can treat them as especially important is because of the role they play – a purely *logical* role, rather than any physical one – in systematising all the other truths.

How, then, does this account need to be modified to incorporate prob-abilistic laws? In the original set of sentences – the set that contained all contingent truths, completely unsystematised and pruned down – there would have been sentences like 'This particular atom of radium-226 decayed 312.045 years after its creation.' There would have been many such sentences, and they would have described a bewildering array of periods of time that it takes for the 226 isotope of radium to decay. What sort of sentence in the messy set could we select as a good way of systematising this phenomenon of radium decay? It would not be appropriate to select a sentence like 'All atoms of radium-226 decay within 3,200 years of their creation.' That would be inappropriate for two reasons: first, such a sentence is probably *false*. As it happens, only about three-quarters of a sample of radium-226 could be expected to decay in that period of time. Second, even if we increased the number of years dramatically, such that it was indeed true that *all* atoms of radium-226 decayed in that period of time, the sentence we would get as a result would be relatively uninformative. A sentence such as:

(1) All atoms of radium-226 decay within 10 billion years of their creation,

for instance, might be true, but does not help us to predict with much accuracy when individual atoms will decay.

Lewis's suggestion is to allow ourselves *other* statements, that are not in the original set of categorical truths, to be included in the best system. These statements are probabilistic statements. A sentence like 'The half-life of radium-226 is 1,602 years' is precisely the sort of probabilistic statement that we might want to include. To rephrase this statement so that it explicitly mentions probabilities, it says:

(2) In any 1,602-year period, an atom of radium-226 has a 0.5 probability of decaying.

This sentence looks relatively simple, and it is much more informative than sentence (1) above. For instance, by using a sentence like this, I can estimate quite precisely (though only probabilistically) how much of a sample of radium will remain after a given period of time.

The problem is, *which* probability statement are we to choose to include in the set? Why pick the statement that ascribes a half-life of 1,602 years, rather than some other period? You cannot say that it is because other statements are obviously *false*. This is a general point about probability statements: unless they ascribe the extreme values of zero or one to an event, then they are strictly compatible with *any* pattern of observations.

So we do not pick the probabilistic statements to admit into the system on the basis of whether or not they are true or false – we can't say at this stage which ones are true or false. Rather, we select them on the basis that they are simple, strong, and *make the actual pattern of events highly likely*. Lewis calls this final criterion 'fit'. Ascribing a half-life of 1,602 years to radium-226 makes it come out as likely that we will observe radium decaying in the fashion that it does. Ascribing a much larger or much smaller number will make it come out as far less likely. That is, the half-life of 1,602 years has a high degree of fit. Having done this – if there is a set of sentences which optimally balances strength, simplicity, and fit – then the *truth* about objective probabilities is given by these sentences.

Again, note that the truths about chances are not made true by some metaphysical role or status. Rather they earn their status as the truths about chance in virtue of a logical-cum-probabilistic relationship that they bear to other truths.

The notion of 'fit'

It is worth spelling out in a little more detail what is involved in obtaining a good fit between a chance theory and actual history.

Suppose that I toss a coin four times, and I obtain two heads and two tails: HTTH. If I adopt the chance hypothesis that the probability of heads on any given toss is 0.5, and the same for tails, then what is the likelihood of *this particular sequence*, given that chance hypothesis? It is

$\frac{1}{2} \times \frac{1}{2} \times \frac{1}{2} \times \frac{1}{2} = \frac{1}{16}$. Students of probability are sometimes inclined to forget that the chance of *any* particular sequence of chance events is almost inevitably very low. Indeed, the chance of this sequence is just the same as a sequence in which the tosses come up all heads, or in which they come up all tails. (What we *can* assign a relatively high probability to is the proposition that, *regardless of the sequence, the frequency of heads will be about 0.5*. For instance, in this case, the chance that the frequency of heads will be precisely two out of four is $\frac{1}{2}$. The chance of getting zero heads is merely $\frac{1}{16}$.)

Despite the probability of this sequence of events being prima facie low, the chance hypothesis we have adopted gives a higher fit than alternatives, such as the hypothesis that the chance of heads is $\frac{1}{10}$ and the chance of tails is $\frac{9}{10}$. On that hypothesis, the chance of this sequence is $\frac{1}{10} \times \frac{9}{10} \times \frac{9}{10} \times \frac{1}{10} = \frac{81}{10,000}$, which is less than one in one hundred – much smaller than $\frac{1}{16}$. In general, chance hypotheses which match the frequencies will have a higher degree of fit than chance hypotheses which diverge from the frequencies. So the requirement of fit will, in large collections of trials, require that the chances be relatively close to the frequencies.

In comparing chance theories for fit with actual history, then, we are comparing relatively small probabilities. For increasingly large sequences of chance events, the probability that they will turn out in any one particular way is going to get ever smaller. But although the probabilities may be small, some chance hypotheses will fit the actual history better than others.

But what if the sequences of events that we are trying to fit to a chance hypothesis are not merely large, but are infinitely large? In that case, as Adam Elga (2004) has argued, this will lead to large numbers of chance hypotheses having *zero* fit – including chance hypotheses which we intuitively would have thought to be good candidates for truth. So the criterion of fit won't do any useful work in ranking alternative chance hypotheses.

To see how this problem arises, imagine that there are infinitely many coin tosses in our world. The sequence begins HHHTH and then is followed by infinitely many landing heads and infinitely many landing tails. Consider the chance hypothesis which assigns chance $\frac{1}{2}$ to getting heads and the same chance to tails. The likelihood of this particular sequence of events is then: $\frac{1}{2} \times \frac{1}{2} \times \frac{1}{2} \times \frac{1}{2} \times \frac{1}{2} \times \ldots$ and so on. The value of this series is zero.

Now consider a different chance hypothesis: the hypothesis that the chance of heads is $\frac{1}{10}$, and the chance of tails $\frac{9}{10}$. The likelihood of this particular sequence, given these chance hypotheses, is equal to:

$$\frac{1}{10} \times \frac{1}{10} \times \cdots \times \frac{9}{10} \times \frac{9}{10} \times \cdots = 0.$$

So both chance hypotheses assign a likelihood of zero to this particular sequence of outcomes. And this will be the case with any infinitely long sequence of outcomes.

There are more sophisticated mathematical techniques that can be deployed here, such as appealing to infinitesimal numbers: numbers that are smaller than any real number. But, as Elga argues, these techniques cannot recover plausible outcomes in this case.

Elga proposes that we look, not for high likelihood of the *particular sequence*, but for *properties* of the sequence that are what we might think of as 'typical' properties, given the particular chance hypothesis we are considering. So, for instance, on the hypothesis that the chance of heads is $\frac{1}{2}$, we would typically expect that the *limiting frequency* of heads would be $\frac{1}{2}$. We would also expect that there would be no obvious *patterns* in the sequence, such as HHTTHHTT ad infinitum. We would expect that, if we just focused on every *second* trial, we would similarly see no patterns, and we would see the same limiting frequency. Elga suggests we call properties like these the 'probability profile' of a sequence of trials. In choosing a chance hypothesis that fits an infinite sequence of data, we should adopt a hypothesis that *gives high probability to the probability profile that the sequence has.*

This technique shows promise, but as it stands we have been quite vague about what a probability profile is: I simply gestured at examples, without giving a recipe for generating an exact profile. And if we are to provide such a recipe, note that we inherit something like the reference class problem again: in choosing the properties that we think of as typical, we are again choosing something like the *relevant classes of events.* See Williams (2008: 409 n.) for further discussion of possible responses to this concern.

Problems for sophisticated actualism

A number of problems arise for a sophisticated actualist theory of chance, and in many ways they are just variations on the same basic idea that I called the 'fundamental' problem for simple actualism. In attempting to replace the concept of a chance with something that is entirely reducible to actual events, the actualist has deviated from the original chance concept. Although the divergence is less stark in this case, it remains a concern that the actualist substitute is not satisfactory.

First, there is a problem for the Humean regarding the likelihood of events in counterfactual scenarios. Suppose I have learned that the half-life of radium-226 is 1,602 years. I take a sample of it, and I leave it for that period of time to see how much decays. Almost precisely half of it does indeed decay. Of course, I know that it is *possible* for it to decay faster or slower than that. It might have been that, after 1,602 years, absolutely *all* of the sample was still intact. I believe this to be spectacularly unlikely, but possible.

How probable is it? Well, to perform the calculation, I would need to know how many atoms of radium are in the sample. For each atom in the sample, it had a 0.5 probability of decaying, and a 0.5 probability of not decaying. So the probability that, in a collection of n atoms, none of them decay is 0.5^n. Supposing the sample to be reasonably big – 10 grams – then the probability would be absolutely tiny – substantially less than 1 chance in 10 raised to the power of 100.[12]

So we can say that, although the sample might not have decayed at all for 1,602 years, we are confident that such a process would be *very unlikely*.

Lewis's theory, however, has a strange consequence here. It says that, although that possibility is unlikely relative to our world, it may be much more likely in the world where it happens! To see this, imagine yourself now in the possible world where the radium does not decay. Moreover, suppose that this is *all the radium there is*. There are no other samples of radium in the world. In that case, if we applied the best-systems analysis to the world, we would not put a sentence like 'The half-life of radium-226 is 1,602 years' into the best systematisation. We would not, because to do so would give us a system of sentences that fares poorly on the criterion of fit. It would make the events in this world highly unlikely. So the half-life of radium-226, in this world, would be different from what it is in the actual world.

That is very odd. Notice how we would normally speak of the possibility of the radium not decaying. We would say something like:

---

12 To work out the probability, it is simply 0.5 raised to the power of the number of atoms in the sample. There are approximately 2.66×10^{22} atoms in 10 grams of radium-226, and 0.5 raised to the power of such a spectacularly large number is – unsurprisingly – spectacularly small.

If the radium were not to decay at all for 1,602 years, that would be extremely unlikely (because the half-life is 1,602 years).

We would usually be wrong to say something like:

If the radium were not to decay at all for 1,602 years, that would be more likely than you think, because the half-life would be different from what it actually is.

But the Humean reductionist seems committed to the second of these being correct.[13]

This is part of a more general issue: for the Humean, by substituting for our concept of chance something that is entirely dependent upon actual events, rather than upon facts about possibilities, the way chances feature in explanations will appear incorrect. We want to use chances to explain why events that occur are likely or unlikely. 'That event was unlikely, because the chance was low.' But for the Humean, one ultimately explains chances by reference to what occurs. So a Humean should endorse explanations such as: 'That event was very unlikely, because not many similar events happen; therefore the chance is low.' This seemingly gets the direction of explanation the wrong way around.

Further issues arise from the use of a notion of *simplicity* in the best-system analysis. First, note how important simplicity is to make the best-system analysis yield non-trivial chances. If there were no constraint on simplicity, then there would be no reason to include probabilistic statements at all: we could achieve maximum strength and likelihood by retaining our original set of sentences.

Here is Lewis on the way in which simplicity might contribute to determining the chances:

In the simplest case, the best-system analysis reduces to frequentism. Suppose that all chance events . . . fall into one large and homogeneous class. To fall silent about the chances of these events would cost too much in strength. To subdivide the class and assign different chances to different cases would cost too much in simplicity. To assign equal single-case chances that differed from the actual frequency of the outcomes would cost too much in fit. For we get the best fit by equating the chances to the frequency; and the larger the class is, the more decisively is this so.

But suppose the class is not so very large; and suppose the frequency is very close to some especially simple value – say, 50–50. Then the system that assigns

---

13 There are some further moves open to the Humean here. For instance, counterfactual sentences seem to display significant context-sensitivity. Perhaps both of these counterfactuals can be true, in different contexts, for the Humean. This might be less implausible than my original conclusion, but it still seems problematic. Why should there be *any* context in which the second counterfactual is true?

uniform chances of 50 per cent exactly gains in simplicity at not too much cost in fit. The decisive front-runner might therefore be a system that rounds off the actual frequency.   (Lewis 1994: 481)

Lewis is imagining, I take it, something like the following scenario: in all of world history, there might only be 846 events of a given sort, C. Suppose that 421 of these events turn out to cause phenomenon E, and no other factors can be cited in a simple generalisation to explain which C-events cause E-events. So the frequency with which C causes E is $^{421}/_{846}$ – just under 50 per cent (approximately 0.49764). The hypothesis that C-events have a 0.5 chance of causing E-events is *simpler* than the hypothesis that C-events have a $^{421}/_{846}$ chance of causing E-events, though it has a marginally less good fit. Nonetheless, the gain in simplicity is sufficient that the 0.5 chance hypothesis makes it into the best system, and is thereby definitive of the chance. Note that, even if this application of Lewis's idea is misplaced in this particular case, the point is that the demand of simplicity is strong enough that it at least *sometimes* has this consequence: otherwise we are left with a view that is very close to simple frequentism.

This story has much to recommend it as a description of our *epistemology* of chance. Of course, if faced with a situation such as this, we would be very likely to adopt the 0.5 chance hypothesis rather than the finicky hypothesis that the chance that C causes E is approximately 0.49764. But why should we suppose that our preference for a simple hypothesis is always going to be reflected in the world? To do so smacks of an excessive optimism about our epistemic powers.

Notice just how strong the implication is, then, of Lewis's view. It is *impossible* for there to be a world just like the one described, in which the frequency with which C causes E is $^{421}/_{846}$ and for the chance to be anything but 0.5. *It is not even possible for the chance to match the frequency.* This seems absurd. Our naive concept of chance implies that non-trivial chances are compatible with every possible frequency. Lewis's theory – like all frequentist theories – is to some extent intolerant of deviant frequencies. Consequently, Lewis's theory appears to reflect an implausible assumption that chances will only exist where they are conveniently close to the frequencies and conveniently simple.

33 Can actualism explain the normative role of chance?

Actualist accounts of chance *must* be revisionary. Because the ordinary concept of chance allows a massive divergence between what actually occurs and the chances, none of the actualist accounts – all of which tie what is objectively probable to what actually happens – can avoid overturning some of our usual assumptions about chance. This defect in actualist accounts, however, might be forgiven, if actualists were able to give a good explanation of what I take to be the central role of chance: its role in prescribing rational belief. We saw, in the previous chapter, that possibilist accounts struggle to explain this feature of chance. Perhaps, because actualist accounts are grounded in what actually happens, we can use such an account to give a better explanation of why we should believe in accordance with the chances.

One defender of a Lewisian understanding of chance, Barry Loewer, has tackled this issue as follows:

Recall that a Best System is one that best combines simplicity, informativeness, and fit. The fit of a theory is measured by the probability that it assigns to true propositions. Fit can be understood as a kind of *informativeness* . . . The higher the probability assigned to true propositions, the more informative the theory (the higher the probability to false propositions the less informative and the more misleading the theory). But these probabilities are informative only to someone who is willing to let them constrain her degrees of belief. Now, suppose that someone decides to let the Principal Principle constrain her degrees of belief. Then for her the Best Theory will be one that best combines simplicity and informativeness – including informativeness as evaluated in terms of the degrees of belief she assigns to true propositions. On this proposal the Principal Principle is 'built into' the account of chance. It can constrain belief because that is part of the account of how a theory earns its title as 'Best System'.

The preceding considerations show that a Humean who values possessing the Best Theory of the world . . . and whose Best Theory contains probability statements must adopt some principle that extracts information about the Humean mosaic from these statements. The Principal Principle is a way of doing that. In the process it provides an interpretation (i.e. truth conditions) for probability statements. So a Humean who values possessing the Best Theory has a reason to adopt the Principal Principle. (Loewer 2004: 1122–3)

Note how Loewer's rationale turns essentially on the attitude of *valuing* one's possession of 'the Best Theory of the world'. This calls for some careful consideration. Consider three possible ways one could value theories of this sort:

1. One could value having a theory because it *maximally combines truth and strength* (the non-probabilistic form of informativeness).

2. One could value having a theory that optimally combines simplicity, strength, and fit (the Lewisian 'Best Theory' criteria), but have this preference for *no particular reason*.

3. One could value having a theory that optimally combines simplicity, strength, and fit (the Lewisian 'Best Theory' criteria) because it is *maximally useful*.

The first understanding of what one values makes a good deal of sense: truth and strength are clearly epistemic virtues of a theory. But by ignoring the requirement of simplicity, this attitude would lead to a 'best system' that eliminates non-trivial chance. The maximally strong and true theory contains no probabilistic statements, but simply the enumeration of all contingent truths. Loewer owes us some account of why we would not choose this as our 'Best' theory.

The second option *fetishises* the Lewisian criteria for being 'best'. It does not answer our request for an explanation of *why* we should forgo the maximally strong and true theory: it simply stipulates that we do. But given that simplicity is not an obvious epistemic virtue, where truth and strength are, this is inadequate.

So we are led to the third option: we value theories that are less than maximally strong because they are more useful *to us*. Maximally strong theories are unwieldy: they may be infinitely complex; they are clearly of little use to limited creatures like us. But note that this rationale explains why using probabilistic hypotheses might be a useful *heuristic*, a technique or practice that benefits us, given our peculiar epistemic limitations. It does not give any reason to think that there are human-independent chance facts that underwrite a human-independent norm such that any epistemic agent should conform his or her credences to the chances. At bottom, the actualist approach championed by Lewis and Loewer is committed to a deep anthropocentrism about chance, which is implausible when presented as part of a realist account.[14]

14 Carl Hoefer (2007) has also defended an elaboration of Lewisian ideas about chance, which is vulnerable to similar objections. In explaining why we should set our beliefs to match the chances, Hoefer says that his view can be characterised as differing from simple frequentism in the following ways, among others: (i) that the distribution of outcomes of a process that we deem chancy must 'look chancy' in the appropriate way; and (ii) that the notion of chance has been 'anchored' to 'our epistemic needs and capabilities through the Best Systems aspect of the account' (2007: 581). Both of these features of his account embody reference to our peculiar epistemic limitations. A more powerful intelligence than ours might find that a distribution of outcomes does *not* 'look chancy'. A more powerful intelligence than ours might be able to use a

In light of the failure of actualism to deliver a plausible realist account, we now turn to accounts of chance that abandon realism.

---

best system that gives a much less central role to simplicity, and thus gives a higher priority to the criteria of fit and strength. In essence, Hoefer's justification of the Principal Principle is a justification that is *addressed only to us*. Of course, that is not nothing: Hoefer and Loewer have made clearer than Lewis did why chances might be useful. But the question that lingers afterwards is: is it really plausible that chances, so characterised, are a mind-independent feature of the world? Are chances *real physical quantities*? Or are chances merely optimal compromises – ways for *us* to marshal our epistemic faculties against a world of great complexity?

8 | Anti-realist theories of chance

In this chapter, I examine a number of ways in which one can take a less realist attitude towards chance than I have been adopting thus far. It turns out that there are at least three important varieties of anti-realism, and it pays to distinguish between them carefully.

34 Varieties of anti-realism

To this point, I have assumed some form of objectivism about chance. By this, I mean that the sorts of facts that make chance ascriptions true are – by and large – facts that obtain independently of our beliefs and attitudes.[1] An alternative, *subjectivist* view is that the truth about chances depends in some way on subjective facts, such as what we believe, desire, or expect. That I like dark chocolate more than other varieties is a subjective fact. Prima facie, that the coin is fair, and thus has an even chance of landing heads or tails, is not a subjective fact, because this fact about the coin surely does not depend on anything to do with anyone's *attitude* to the coin. But perhaps that appearance is mistaken, and subjectivism might provide a better approach to understanding chance.

Both objectivism and subjectivism assume that chance claims represent facts. But this assumption too can be denied. Some meaningful sentences seem to have meaning without representing facts. For instance, commands and exhortations such as 'Shut the door!' and 'Please help me' do not seem to represent how the world is – at least not as their primary meaning. Similarly, exclamations such as 'Ouch!', 'Hooray!', and 'Hmph' seem to be expressions of our attitudes, but they are not themselves *representations* of our attitudes that are apt to be true or false. These areas of language seem to be meaningful in virtue of performing functions other than representing the

---

1 The possibility that chance ascriptions might be contextually sensitive complicates this a bit. One widely accepted way of thinking about context-sensitivity is that, although it might mean that *what proposition is asserted* by a given sentence may depend, in part, on subjective factors, the truth conditions for the proposition asserted need not depend on subjective factors. So I take the sort of context-sensitivity discussed in Chapter 1 to be compatible with objectivism.

world. Commands and exhortations seem to function by prescribing certain behaviour for others to perform. Exclamations function by expressing our mental states, without saying that we have those mental states.

I have been assuming that chance claims do represent facts. This is part of what I mean by the idea that chances are *real*. An alternative, *non-cognitivist* approach to chance is one that takes the meaning of chance claims to be in some sense non-representational. Chance claims are not attempts to represent facts – be they objective or subjective. Rather, they are better understood on the model of some other, non-representational part of language such as exclamations or commands.

Both subjectivism and non-cognitivism are forms of anti-realism that involve semantic claims – they are both committed to revisionary ideas about the meaning of chance ascriptions. A third form of anti-realism is more conservative in semantics: it is possible to concur with the objectivist thesis that what makes chance ascriptions true is entirely a matter of objective facts. But one could then deny that *any* chance claims are true. This form of anti-realism is known as *error theory*. A decision tree, explaining the relationships between the different forms of anti-realism, is shown in Figure 8.1.

The task of this chapter is to try to identify the principal benefits and disadvantages of adopting an anti-realist approach to chance. I begin with the semantically conservative form of anti-realism: error theory.

35 An error theory of chance

An error theorist is an anti-realist about chance in the way that most of us are anti-realists about ghosts or vampires. Talk of ghosts and vampires functions much like talk of cats and dogs. Sentences like 'There is a ghost in the attic' purport to represent a mind-independent state of affairs. Such a sentence *could* be true, in much the same way that 'There is a dog in the yard' could be true. But if there are no ghosts or other supernatural creatures, then all such talk of ghosts is false, or at best is true in a non-literal sense.[2]

What reasons could move one to think, in this way, that chances simply do not exist?

First and foremost, one might be impressed by the failure of the attempts, discussed in the previous two chapters, to provide a realist account of

---

2 A sentence like 'Banquo's ghost is covered in blood' is quite acceptable, but it is doubtful that it is literally true. Better, it should be understood as tacitly qualified by an operator such as: 'It is true in *Macbeth* that . . .'

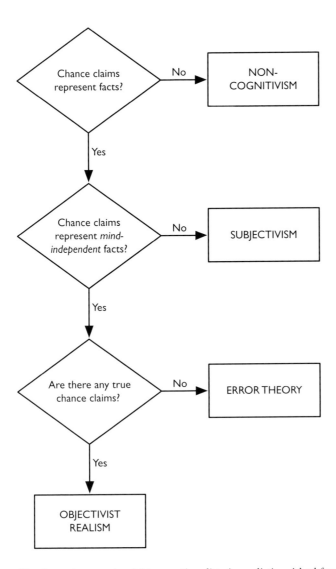

Figure 8.1 The three views on the right are anti-realist views, distinguished from each other primarily by their semantic commitments: each anti-realist view understands the meaning of chance ascriptions differently. Only error theory shares the objectivist semantics of realism.

chance in either actualist or possibilist terms. If no one has yet provided a satisfactory realist theory, perhaps that is because there is nothing to provide a realist theory of.

Second, we can reflect more generally on the sort of phenomenon that chance must be, if it is to meet the various precepts and platitudes that we associate with chances: chances are physical quantities; chances are not reducible to actual frequencies; and yet our beliefs about chances stand in a normative relationship to our expectations about frequencies. This is a difficult combination to pull off. Perhaps it is impossible for anything to meet these desiderata, so there can be no chance – at least not as classically conceived.

But, as it stands, this is not a very strong basis for adopting error theory. Comparison with the case of ghosts is instructive. Claims about ghosts have never provided good explanations for empirical phenomena. People who reject all claims about ghosts do not do noticeably worse in everyday life than those who affirm a belief in ghosts. But chance claims are part and parcel of our ordinary dealings with the world. Doctors recommend surgical procedures or medications on the basis of their estimates of the chances. Professional gamblers get rich or poor to the extent that they are able to choose to play games where the chances of winning are greater than they are of losing. Quantum physicists build exquisitely sensitive experimental apparatus to test chance hypotheses, and often find that their theories are well confirmed.

More generally, claims about chance have been part of our way of understanding the world for centuries. We have had explicitly chancy scientific theories at least since quantum mechanics, but almost certainly before then. Some chance theories have been defeated, while others have been confirmed. It is very plausible to believe that well-confirmed scientific theories are approximately true. So it is plausible that our well-confirmed chance claims are approximately true.

Admittedly, this critique of error theory is vulnerable in two ways. First, it assumes scientific realism, which is a surprisingly difficult doctrine to defend. Much as it comports with common sense to think that successful scientific theories track the truth, it is very difficult to justify the extension of this attitude beyond claims about the observable realm to sentences that posit more abstract and theoretical entities such as causal relations, numbers, and chances. Second, even if scientific realism is correct, it only establishes the *approximate* truth of chance claims. Perhaps the true claims in the vicinity are not, strictly speaking, chance claims but are some not-quite-chancy surrogates.

Despite these vulnerabilities in scientific realism, it is fair to say that error theory lacks much credibility until something is provided by way of a 'palliative' account. If all chance claims are false, why are they so useful? If all chance claims are false, why have so many of us taken so long to notice? Instead of supposing that all chance claims are false, might it not be more plausible to suggest that chance claims do not function in the same way as other scientific claims, and thus avoid the need to provide a representational, mind-independent account of their meaning? And this question, of course, leads us to the other varieties of anti-realism.

36 Subjectivist interpretations of chance

There is a reasonably clear connection between the chance *claims* we make and the sorts of *beliefs* we have – or at least the beliefs we ought to have. For instance, if you assert that the chance that we will discover life on another planet is high, and I accept this claim, then I should – in some sense – lend a higher degree of belief to the proposition 'We will discover life on another planet.' It would be very strange if I granted you that the chance of it being true is very high, but insisted that I had no inclination whatsoever to believe it.

Sometimes this link between chance and degree of belief will come apart, but typically this will be evidence of irrationality on the part of a person. A gambler, for instance, might acknowledge that the chance of a win on the next spin of the roulette wheel is only one in thirty-seven, but maintain a high degree of confidence that he will win, because of a hunch that this is somehow a 'special moment'. Putting such cases aside as non-standard, the subjectivist approach tries to use this connection between chance and degree of belief to give a *definition* of chance. All probabilities – including chances – are simply degrees of belief. There might be further special conditions that demarcate the chances from other subjective probabilities, but chances are, at root, credences.

As it happens, it is very hard to find anyone who explicitly subscribes to this view. There are a number of writers who have emphasised subjective elements in writing about chance, but it is hard to find them owning up to the thought that we can actually analyse a chance ascription in terms of subjective states of mind. That is, a subjectivist should be committed to something like the following analysis:

s. A chance ascription 'The chance of E is x', uttered by S is true if and only if . . .

and the right-hand side of this analysis should mention something about the subjective states of S. A particularly simple form of subjectivism would be:

s1. A chance ascription 'The chance of E is x', uttered by S is true if and only if S's degree of belief in E is x.

A moment's reflection reveals a number of reasons why any such analysis is very implausible. Suppose that Tweedledee and Tweedledum are discussing the likelihood that the Cheshire Cat will be grinning once it realises they have stolen its supper.

Dee. It will be smiling away like mad. At least, that's what's most likely.
Dum. Hardly! I concede there is *some* chance it will keep its grin, but it is far more likely that it will be frowning.

Translating the chance talk along the lines of (s1), this dialogue would amount to:

Dee. It will be smiling away like mad. At least, I believe more strongly that it will be smiling than I believe the contrary.
Dum. Hardly. I have some degree of belief that it will retain its grin, but I have a stronger degree of belief that it will be frowning.

The problem with this translation is that Dee and Dum are no longer disagreeing – at least not in their chance talk. Rather, they are just reporting to each other their different subjective probabilities. To the extent that they think they are disagreeing about probabilities, they are actually *talking past* one another.

So subjectivism is off to a poor start. However, those working in a tradition of probabilistic *epistemology* that is known as 'subjectivist' have made a number of observations about the psychology of chance belief which take us a long way from an analysis as crude as (s1). Although it is not always clear what the metaphysical import of these theories is, we can explore this programme as a means of fleshing out a more defensible subjectivist theory. And even if we cannot use these insights to resurrect subjectivism itself, they may be of more use when we turn to non-cognitivist approaches.

The subjectivist theory denies that there are objective facts to which probability claims must correspond in order to be correct. If you believe that the probability of my cat, Tibbles, being elected prime minister of Australia is very high, and I believe the probability to be very low, then there is nothing about the *world* as such that makes my belief correct and yours false. That is just a difference in our attitudes.

This might seem objectionable, because surely there is something *criticisable* about you for believing that Tibbles will probably become prime minister of Australia. But if there are no facts about probabilities, then I cannot criticise you on the grounds that you have a false belief. I seem to lack any basis for preferring my state of mind to yours as the better warranted state of mind.

The subjectivist does, however, have some basis for criticising at least *some* agents who assign strange probabilities to events: a rational agent needs to follow certain rules of probabilistic consistency in her beliefs – both at a particular time and across time. Failure to do this – claims the subjectivist – will lead to a probabilistic form of incoherence. One will have beliefs about probabilities that do not make collective sense.[3] We have already examined, in §3, the sort of consistency that is required at any single time. For instance, for all propositions, one must ascribe credences no less than zero and no greater than one. And for logical compounds of propositions, there must be certain relations between the credences in the compounds and the credences in the propositions from which they are composed. For example, our credence in a disjunction, $Cr(P \text{ or } Q)$, cannot be lower than our credence in either of the disjuncts: $Cr(P)$ and $Cr(Q)$.

Across time, the sort of consistency required is that of updating one's credences in accordance with the new evidence one obtains. Suppose you had not known that Tibbles was not born in Australia. Moreover, you think that a cat's being born outside of Australia is good evidence that that cat will not be elected prime minister. Your credences, then, might be as follows:

1. Cr(Tibbles will be elected prime minister) is high.
2. Cr(Tibbles will be elected prime minister, given Tibbles was not born in Australia) is low.
3. Cr(Tibbles was not born in Australia) is low.

But now some new evidence comes in. And this evidence changes the third of these credences. The new evidence increases your confidence in the proposition that Tibbles was not born in Australia. So your new credence becomes:

3′. Cr(Tibbles was not born in Australia) is very high.

The combination of credences (1), (2), and (3′) is incoherent. You need to change at least one of those credences in order to regain coherence. A natural thought is that you should modify (1), the credence that Tibbles

---

3 Bruno de Finetti, a notable subjectivist, wrote in this vein:
    In investigating the reasonableness of our own modes of thought and behaviour under uncertainty, all we require, and all that we are reasonably entitled to, is consistency among these beliefs, and their reasonable relation to any kind of relevant objective data ('relevant' in as much as subjectively deemed to be so).   (de Finetti 1974: x)

will become prime minister. Your unconditional credence that my cat will become prime minister should adopt a similar value as you currently have for the *conditional* credence, (2). Because you are nearly certain that Tibbles was not born in Australia, then the probability that Tibbles will be elected PM is going to be similar to the conditional probability that Tibbles will be elected PM, given Tibbles was not born in Australia. This technique of revising one's credences is known as *updating by conditionalisation.*

It is certainly demonstrable that this is *one* coherent way of updating your credences.[4] But it is doubtful that this is the *only* coherent way of updating your credences. The axioms of probability theory tell us that the combination of (1), (2), and (3′) is incoherent. But the axioms do not tell us a preferred way to rectify the inconsistency.

And there are other options. You could simply abandon your conditional credence (2). This might involve saying to yourself something like: 'Oh! Since I am sure that Tibbles will be PM, but I now also think that Tibbles was not born in Australia, I must have been wrong to think that being born outside Australia makes it unlikely that Tibbles would be PM.'

Or you could simply abandon what you took yourself to have learned about the origins of my cat, by rejecting (3′). 'I know this feline birth certificate looks authentic, but it *must* be a fake, because I am sure that Tibbles will be PM, and if Tibbles were born outside Australia, he would most likely not become PM.' These might sound like irrational responses, but using the minimal resources of the probability calculus, it is not easy for the subjectivist to show that these alternative responses are improper.

So, while it is not true that subjectivists have no means of criticising agents with eccentric probabilistic beliefs, they are somewhat limited in the means of criticism they can employ. Although we can all agree that simultaneously holding the attitudes described by (1), (2), and (3′) is to be probabilistically inconsistent, realists about chance can say that we should avoid credences that don't match the chances. The subjectivist cannot say this. For the subjectivist, the only standard of criticism is coherence. Therefore there is no objective standard by which to say which of the above three credences should be rejected.

Another problem for subjectivists is that our chance claims feature in physical explanations. For instance, we can use the half-life of an isotope to

---

4 The arguments used are *diachronic* versions of the Dutch Book argument, which was discussed in §3. David Lewis is credited with the first form of the argument: Lewis (1999: 403–7); Teller (1973). See Christensen (1991) for some doubts about the provability of this point and Briggs (2009c) for some counter-moves in defence of the diachronic Dutch Book argument.

explain what happens to a sample of that isotope. It is very hard to see how purely subjective probabilities can play this explanatory role. Imagine two interlocutors discussing some plutonium that is decaying:

Look, half the plutonium has gone. Why's that?

Oh that has happened because it obeys a law of decay such that half of it goes every hour or so.

Is that a deterministic law?

No, it's just a probabilistic law. That is, there is a chance of 0.5 that any given atom will decay in any given one-hour period.

What do you mean by saying there is a 'chance'?

Being a subjectivist, I mean that my personal degree of belief that a given atom will decay in any period of duration one hour is 0.5.

But you were trying to explain to me why it has decayed. Are you saying half of it has decayed in an hour because you *believed* it would decay?!

What we seem to get here is a form of explanation that explains the wrong thing: it explains why the agent who believed in the probabilistic law is not *surprised* by the decay. But we don't want an explanation of the mental state of the observers: the idea is that the law should explain what *actually happens*.[5]

Another point worth mentioning about these sorts of chances, as they arise in scientific explanation, is that the chances appear to be *intrinsic* properties of the experimental set-up. The chance of decay depends upon the nature of the atom, and on what forces it is subjected to. The chance does not depend on the state of mind of the experimenter, or on what anyone else knows about it. But again, because a subjectivist thinks that all probability amounts to subjective degrees of belief, she does not have the means to explain the way in which a physical chance could be an intrinsic property of a physical set-up, independent of psychology.

37 The subjective psychology of objective chance

We often take ourselves to be investigating physical probabilities. That investigation seems to be an investigation of something *mind-independent*. As we saw above, a subjectivist translates all claims about probability into claims

---

5 This objection is fiercely denied by, for instance, Skyrms (1984: 13). Because of this denial, I doubt that it is apt to label Skyrms a 'subjectivist' as I am using the term. Rather, there are hints that Skyrms would be better categorised as a sort of non-cognitivist. Indeed, non-cognitivism looks like the most promising way for an anti-realist to show that chances can play an explanatory role.

about subjective psychology. So it seems impossible for a subjectivist to explain what it is that we are doing when we investigate physical chances.

There is, however, one very impressive result in the tradition of subjectivist epistemology which might be thought to largely address this concern. These theorists have shown that they can give an idealised psychological account of what is involved in investigating what we would ordinarily take to be a matter of 'objective chance'. These accounts have thoroughly respectable subjectivist credentials, in that they do not make reference to any objective states of affairs. Perhaps if we use this account to supplement the very crude analysis (s1), we might get a more plausible result.

The first part of this strategy is best exemplified in some highly technical work by the Italian theorist of probability, Bruno de Finetti. De Finetti has shown that any agent who meets certain psychological requirements regarding some possible experiments is equivalent to an agent who is entertaining 'objective chance' hypotheses. This is known as de Finetti's *representation theorem*. In the following section, we will work through the essentials of this idea.

First stage: utility and credence

First, as we already touched upon in the opening chapter (§3), some subjectivists such as Ramsey and Savage have tried to show that a given credal state is *equivalent* to having a particular pattern of *preferences* over wager-like choices. The claim is controversial, but suppose we grant it.[6]

Second, there is a family of formal proofs, known as 'representation theorems', which show that anyone who possesses a set of preferences that has sufficient structure, and that meets a number of rationality constraints, will behave like a person with three crucial properties. She will appear to (i) have a utility function, (ii) have credences that obey the probability calculus, and (iii) be behaving in a way so as to maximise utility.

The assumptions of representation theorems

This manner of proof is foundational to a branch of probability theory known as 'decision theory', because it offers a formal method of analysing rational decisions. As with most formal approaches to understanding a

---

6 I am ignoring in my presentation of these ideas some important differences between different thinkers in the subjectivist tradition about the exact order of explanatory priority between different sorts of mental state. See Logue (1995: §1.1) for a brief but helpful survey of some of the options.

messy natural phenomenon such as decision-making, one needs to be aware of the idealisations that are involved in the initial assumptions.

While the rationality constraints that are supposed in decision theory have some plausibility when considered in the context of particular examples, it becomes evident that they are psychologically unrealistic when you consider that the decision theorist expects them to hold among *all* our preferences at *all* times.

For example, one of the assumptions is that your preferences are transitive and irreflexive. Here is a simple example to illustrate the sort of oddity that would arise if you violated those assumptions. Suppose you prefer two doughnuts to one doughnut, and you prefer one doughnut to a banana. But when you are given the choice of two doughnuts or a banana, you feel guilty about eating two doughnuts, and prefer the banana. In that case, using the '>' symbol to represent the relation of preferring one thing to another, your preferences seem to have the structure:

Banana > Two doughnuts > One doughnut > Banana

Suppose you were willing to pay a small amount to trade any of these things for another thing that you prefer. In that case, if you started with a banana, you would be willing to pay to trade it for a doughnut. Then you would be willing to pay again to trade that for two doughnuts. Then you would be willing to pay *again* to trade those for a banana. You would be back to where you started, except that you have paid out money at every step! This is the so-called 'money-pump' argument in favour of meeting the decision-theoretic constraints. Certainly, it does seem irrational to have preferences like this. By conforming to the decision-theoretic axioms, we guarantee that we won't be vulnerable to money-pumping of this sort.

But, less plausibly, the decision-theoretic axioms effectively presuppose that all things have *commensurate value*. For instance, suppose you had to compare options such as:

A: having a somewhat risky, but ultimately deeply fulfilling career as an aid worker in foreign countries, though this means that your personal relationships suffer somewhat, and you never have children;

B: having a less exciting, but still reasonably satisfying, career as a schoolteacher, but marrying happily and having healthy children; and

C: having much the same career and family life as in B, but in this scenario you suffer one extra day of illness with a bad cold.

Presumably you prefer B to C, given that B is much the same as C, but involves slightly better personal health. But you might find, even upon close reflection, you do not prefer B to A or vice versa. According to decision theory, if you do not prefer A to B, and you do not prefer B to A, but you prefer B to C, then you are *rationally required* to prefer A to C. That is, you are rationally required to treat the extra day of illness in C as a tie-breaker that makes a crucial difference, not only with respect to the comparison of B and C, but also with respect to A and C.

That seems rather odd. Surely it would be more natural to think that A is so different from both B and C that it is neither better nor worse than both of them. This is best accounted for by supposing that A is *incommensurate* with both B and C. But decision theory assumes that there are no incommensurate values. (The notion of incommensurate value is helpfully introduced in Raz (1985). See also Broome (2004: §12.1).)

So decision theory oversimplifies. It is not a descriptively adequate theory of rational preferences for creatures like us. But having noted this sort of limitation, it would be too hasty to ignore decision theory altogether. It would be especially hasty in this context since, using this apparatus, we can perhaps obtain a promising way of understanding chance that avoids many of the 'metaphysical' problems encountered so far.

A *utility function* is a function that assigns a value to possible states of affairs that the agent might bring about through her actions. In particular, it is a *cardinal* function, which means that we can make meaningful quantitative comparisons between things that are valued by the utility function. We can say, for instance, that gaining two doughnuts is *twice as good* as gaining one doughnut.

And the idea of being a *utility maximiser* is simply that one attempts, in one's actions, to choose actions which have the highest 'expected utility'. Each action we perform has numerous possible outcomes. Walking to the local shop to buy some milk might achieve the intended outcome and nothing else. But there is also a small probability that I will get there only to find there is no milk. There is an even smaller probability that I will be hit by a car on the way, and require surgery. And so on. Each of these outcomes, if spelled out in sufficient detail, will have an associated utility, and an associated probability that such an outcome will come about. The expected utility of going to the shop is the average of the utilities of all the possible outcomes, weighted by their probability.

So to sum up the programme so far: provided you have a set of preferences over choices that involve an element of uncertainty, and provided those preferences have a rich enough structure, and provided you meet certain somewhat idealised rationality constraints, then it can be *proven* that there is a way to describe your psychology in terms of probabilistic credences and a utility function.[7]

Despite the highly idealised assumptions that are required for the proof, it is nonetheless a profound result. While I have glossed over the technical details, it is worth stressing the sort of conceptual gap that this proof has apparently bridged. David Hume famously argued that beliefs alone were never sufficient to motivate us to behaviour. Hume's idea has been extremely influential, and consequently beliefs and desires are paradigmatically regarded as utterly distinct elements of our psychology – beliefs are constrained by reason, desires are unconstrained by reason. According to the Humean, there is no way of deriving rationally required beliefs from particular desires, or vice versa. But those in the subjectivist tradition of probability claim that, given an agent with a sufficiently rich *preference set* – something akin to a set of desires – and given that the agent obeys some somewhat demanding, but nonetheless relatively plausible rationality constraints, then *ipso facto* that person has a completely determinate and coherent set of probabilistic *beliefs*.

Of course, this is not sheer magic, once you see the trick involved here. The trick is that the preferences have to discriminate between *wager*-like choices.[8] If someone is *indifferent* between all options that involve *uncertain outcomes*, then we won't be able to infer any probabilistic beliefs. (That's why the preference set has to be sufficiently rich in structure.)

Consider a toy example. Suppose someone:

- prefers ice-cream to cake;
- is indifferent between two wagers: (i) a coin toss, with heads = cake, tails = ice-cream and (ii) a second coin toss, with heads = ice-cream, tails = cake; and

---

7 Savage (1954: chaps. 2–5) is a very influential exponent of these ideas.

8 Arguably, another part of the trick is that the preferences involved are unlike desires as traditionally understood, in that there is always assumed to be a perfect match between what action one performs and what one prefers – they are what is known as 'revealed preferences'. By contrast, we frequently think it is possible to act contrary to one's strongest desires. As such, these putative preference states are sufficient to determine action – making them effectively equivalent to Humean belief–desire pairs.

- prefers one wager: (i) to draw from a barrel of red and black balls, with red = ice-cream, black = cake; over a second wager: (ii) a coin toss with heads = cake, tails = ice-cream.

From information like this, we can infer, e.g.:

1. The agent's degree of belief that the coin will land heads is equal to her degree of belief that it will land tails.
2. Her degree of belief that, if she draws a ball from the barrel it will be red is greater than her degree of belief that if she tosses a coin it will land tails.

And with *enough* structure in the preferences, we can go beyond this to derive precise numerical credences, rather than merely an ordinal ranking of credences.

Second stage: learning from experience

Next, we need to consider how this framework can account for learning from experience. Learning, in this framework, amounts to something like changing one's preferences for wagers, on the basis of the outcomes of earlier wagers.

For instance, suppose I am confident that you are using a biased coin in the wagers you offer me, but I'm unsure if it is biased towards heads or tails. You offer me a bet of $10 on heads, zero otherwise. I fancy my chances, so I prefer to accept the bet. Now make a further supposition that, after this bet, you will offer me a bet of $10 on tails, zero otherwise. Right now, I can have preferences about whether or not I take this second bet, *conditional on the outcome of the first bet*. Roughly, these preferences will be of the form:

- If the coin lands heads on the first toss, don't bet on tails on the second.
- If the coin lands tails on the first toss, do bet on tails on the second.

Having conditional preferences amounts to being in a credal state that is sensitive to future events. Indeed, it amounts to having credences that behave like *conditional* probabilities. These particular conditional preferences are roughly:

- Cr(Coin is biased to tails, given heads on first toss) = A low value
- Cr(Coin is biased to tails, given tails on first toss) = A high value

(Again: with a sufficiently rich set of preferences, we would be able to establish precise numerical credences, rather than merely higher and lower credences.)

Return, then, to the original example of the coin which I believe is biased. Because I suspect bias, I have conditional preferences about what to bet next, after the outcome of the first bet is known. Suppose these conditional preferences are *effective*. That is: I *would* come to prefer taking the second bet, if I were to lose the first. And I *would* come to prefer rejecting the second bet, if I were to win the first. If, in general, I always revise my preferences so as to prefer what my pre-betting preferences told me to, conditional on the outcome, then it can be *proven* that I will be following the epistemic updating strategy known as *updating by conditionalisation.*[9] This is the strategy we have already mentioned above, which is certainly one coherent way of updating our credences in the light of evidence, and has been defended by some as the optimal strategy for updating credences in the light of new evidence. For our purposes, we merely need to bear in mind that updating by conditionalisation involves the eminently attractive idea that we should increase our credence in hypotheses that make the observed data more likely, and decrease our credence in hypotheses that make the observed data less likely.

There is much more of interest in the details of this programme, but here's the moral: the 'textbook' practices of subjectivist epistemology (e.g. Howson and Urbach 1989) whereby we confirm scientific chance hypotheses by looking at data, and then revising our probabilistic beliefs by conditionalising on the data, are – according to the subjectivist tradition – instantiated in any agents who have a few traits:

1. they have sufficiently discriminating preferences over wager-like choices;
2. they do not have certain rational defects in their preferences;
3. they have different preferences for what they prefer in future, dependent upon the outcome of earlier wagers; and
4. they do indeed come to adopt those new preferences in light of past wagers.

Third stage: objective chance

An agent who updates his or her credences by conditionalisation is an agent that is in some sense sensitive to evidence. But there are *different* ways of being sensitive to future evidence and they are not all approaches that

9 For details of this sort of proof, the reader would do well to consult original works by de Finetti (1937; 1974) and Skyrms (1984), though the going can be difficult. A more recent paper, (Greaves and Myrvold 2010), contains a marginally more accessible, though still technically rigorous, presentation.

look like investigations of an objective chance. Suppose you are trying to determine whether a roulette wheel is fair. You spin the wheel many times, and find that the frequency of getting 17 is surprisingly low. As the data continues to come in, you continue to become less and less confident that the wheel is fair. In particular, you come to think it less and less likely that a bet on 17 is likely to win. In terms of conditional preferences, you have a conditional preference that, if the outcome of previous trials reveals 17 to be relatively infrequent, you prefer to bet on 17 *less* than you did before.

In light of this conditional preference we can determine that you have certain conditional probabilistic beliefs. Your conditional credence Cr(Second spin = 17, given First spin $\neq$ 17) is lower than your unconditional credence Cr(Second spin = 17).

It is possible to imagine a less wise gambler than yourself who is convinced that, the longer the wheel is spun without 17 coming up, the more likely it is that the next spin will be 17. This gambler has a conditional preference that favours betting on 17 *more*, if the previous spin was not equal to 17. Even though these attitudes are very different, they can both be part of internally coherent subjective psychologies. Both agents, provided their conditional preferences do not violate the axioms already introduced for the representation of credal states in general, may be perfectly probabilistically coherent. And both agents could be exemplifying the strategy of updating by conditionalisation.

The paradigm of an objective chance, however, is a probability that is the *same* between *intrinsically similar trials*. The first attitude is that of someone who treats the roulette wheel as having an unknown, but fixed, physical probability to land 17. So it is only the first sort of attitude that we want to characterise to show how a subjectivist explains the psychology of someone who believes in objective chance.[10]

The subjectivist has limited resources here: she cannot say that one attitude is correct. But she can at least characterise, in general terms, the sort of psychological state that is typical of someone who is entertaining what we might call a *chance hypothesis* about the data. Suppose we find some

---

10 Trials that display this pattern of probabilistic independence are known as Bernoulli processes. We might think that the idea of chance generalises further, to cover Markov processes for instance, which exhibit probabilistic dependence upon the last trial done: e.g. repeatedly flipping a thumbtack. Whether it lands on its back or on its side is not independent of whether it was flicked from being on its back or on its side. See Diaconis and Freedman (1980); Skyrms (1984: 47–62) for an extension of de Finetti's exchangeability idea to these sorts of probabilities.

data that simply looks like a list of successes and failures in a series of eight trials:

D1. No, Yes, No, No, No, No, No, No.

Based on this data, what do we infer about the probability of a success, if another trial was performed? If we are told that the trial was simply the tossing of a coin, and observing whether or not it landed heads, then we might assume this was just an unlucky set of trials and maintain that the probability of heads is very close to 0.5. Or we might be more directly influenced by the data, and conclude that the probability of heads is merely one in eight, or close thereto. Whatever you are inclined to believe, call that probability *x*.

Now suppose you realise that the data is subject to another interpretation. It could be that the trials were not performed in the order recorded. It could be, for instance, that the actual chronological sequence of trials was:

D2. No, No, No, No, No, No, Yes, No.

If that were the case, would it change your belief about the probability of heads, if another trial were performed? Most likely not. This tendency, to be insensitive to the sequence in which the trials were performed, is characteristic of an agent who believes that the matter in question is governed by a chance that is *invariant* between trials.

Compare, for instance, how we would react if we were told this data related to an experiment in animal behaviour, where we provide a single canary with some seed, and watch to see if it eats it within the next five minutes. Now there is a huge difference in significance between D1 and D2. D1 looks like a case where the bird has tried the seed and decided it does not like it. D2 looks like a case where the bird has only just recently decided to sample the seed, and may well try it again. It is extremely dubious that the behaviour of the canary is well described by a single chance hypothesis that is invariant between trials, so the sequence in which the data occurred is highly relevant to what belief we form about the probability of success on a future trial.

We can define precisely what it is for an agent to be indifferent to the sequence in which the data was collected. De Finetti calls this condition 'exchangeability'. Using this concept, de Finetti proved that any agent who treats data as exchangeable can be represented as having a credal state that is equivalent to entertaining a number of 'chance hypotheses', with different degrees of confidence.

This idea of exchangeability captures what is importantly different between gamblers like us, who think that the chance of 17 on the roulette wheel is the same between trials, and the less wise gambler who thinks that the longer it goes without landing 17, the more likely it is to be 17 on the next spin. If the data in sequences D1 and D2 referred to whether or not the wheel has landed 17 in the most recent spins, then gamblers like us would treat the two sequences as equivalent in how they affect our inclination to bet on 17 in future, whereas the other gambler does not treat them as equivalent. For this gambler, D1 shows great promise: there has not been a 17 for several spins. But if D2 represents the recent spins, then betting on 17 is a much less desirable option: it has come up too recently.

So the attitude that we think of as rational, in dealing with a game like roulette, is captured by exchangeability. It must be stressed that the subjectivist does *not* have an account of why we are 'correct' to have that attitude towards roulette. But at least the subjectivist has explained, in terms of a rich set of conditional preferences, two key elements of our normal chance-related behaviour. The subjectivist has explained the phenomenon of treating chances as invariant between intrinsically similar trials. That is just an upshot of treating data from intrinsically similar trials as exchangeable. And the subjectivist has explained the phenomenon of updating our beliefs about the chances in light of statistical evidence. That is just an upshot of having conditional preferences, and revising our preferences in accordance with those conditional preferences.

Probability and epistemology

In the passages above, I have outlined an important result in the broadly subjectivist tradition of probabilistic epistemology. It is beyond my remit to discuss in any great detail the concerns that have been raised, both against that particular result and against the broader programme. But it might be helpful to some readers to briefly identify the sorts of alternative views that are being discussed in the literature.

Probabilism is the idea that rational degrees of belief must obey the axioms of probability. This is a relatively widely shared idea, but note that the earlier example of a sentence like '"Lucky" will win the lottery' might constitute a counterexample (§2).

Bayesians subscribe both to probabilism and to the idea that we are rationally required to update our credences by conditionalisation.

Bayesians can be further divided into *objective* and *subjective* varieties. Objective Bayesians claim that for any body of evidence there is a uniquely permissible credence distribution (e.g. White 2005; 2010). Subjective Bayesians deny this (e.g. Howson and Urbach 1989: chap. 3). Among subjective Bayesians, a particularly extreme view ('mad dog' Bayesianism) is that the *only* constraints on rational credence are the probability axioms and updating by conditionalisation.

All of these are views about the epistemic constraints under which rational agents operate. None is a thesis about chance per se, though as I discuss below, it is tempting to try to vault from relatively subjectivist views about probabilistic epistemology to an anti-realist view about the nature of chance.

We have been reviewing these ideas from subjectivist epistemology in order to see if they support some sort of subjectivist, or otherwise anti-realist, position about the nature of chance. Here is the sort of argument one might mount, drawing on these ideas.

P1. Coherent probabilistic 'beliefs' (and associated epistemic practices) are identical to coherent patterns of preferences.
P2. Coherent patterns of preferences do not represent objective facts.

Therefore,

C. Coherent probabilistic 'beliefs' (and associated epistemic practices) do not represent objective facts.

This argument, as stated, is subject to all manner of objections, but I take it that it at least captures a profoundly tempting thought arising from the subjectivist tradition in epistemology. If probabilistic beliefs are nothing more than complicated desire-based phenomena, then how can they represent how the world is? Preferences and desires are not even apt to be true or false. You prefer bananas to doughnuts, and I prefer doughnuts to bananas. Is one of these preferences more true than the other? The idea is absurd.

But this argument alone is not likely to persuade someone who is still inclined to believe that there are objective chance facts. For if there were such facts, then perhaps we should simply reject that (P2) holds for all preferences. Perhaps preference sets which correspond to chance beliefs *are* – unlike other preferences – representational: they represent the chances!

What the anti-realist needs, to advance the argument further, is to offer some sort of explanation of why we treat some phenomena as chancy, and why we have such a firm conviction that it is *incorrect* to treat those phenomena otherwise. All we have so far is an illustration that the psychology of someone who believes in objective chances is *compatible* with anti-realism. We have not yet had an explanation of why this sort of psychology is so useful, if the world really does not contain any chances.

Moreover, the particular form of anti-realism with which we have been notionally working – namely, subjectivism – is subject to overwhelming objections, on account of its failure to show how chances can feature in explanation, on account of its failure to fit with the way we can engage in meaningful disagreements about chances, and so on. If anti-realism is going to be a defensible view about chance, it will need to stray into different semantic territory, and abandon the idea that chance claims are simply representational claims about our beliefs or other attitudes.[11]

38 Non-cognitivism

In all earlier chapters, I have been presupposing that chance claims are representational. An associated idea is that they are the sorts of claims that are apt to be analysed in terms of their *truth conditions*. This is a linguistic claim about how such claims get their meaning.

In order to see that this is a controversial assumption, it is necessary to observe the variety of sentences in a natural language, and the different ways in which we might try to account for their meaning. Consider a sentence like 'Put the cat on the mat.' In order to assert this sort of sentence sincerely, the speaker does not need to represent very much at all about the world. It does appear that the speaker is making various *presuppositions*. For instance, a sincere speaker presumably presupposes that there exists both a particular cat and a particular mat; and that the said cat is not currently on the mat. But suppose that the cat is already on the mat. One of the presuppositions is incorrect, but we would not say that the speaker has uttered a falsehood. The

---

11 Jackson and Pettit (1998: 250–1) claim that the objection from disagreement, if it succeeds, can be turned equally against non-cognitivism. For non-cognitivists cannot characterise disagreement in the relevant domain as the assertion of propositions which cannot be true together, since they deny that assertions in that domain express truth-conditional propositions. So whatever disagreement means, it might be possible for subjectivists to co-opt it also.

utterance has malfunctioned in some way, but *not* because it represented the world inaccurately.

In contrast, a sentence like 'The cat is not currently on the mat' clearly *does* represent the world as containing a particular cat that is not currently on the appropriate mat. If the cat is already on the mat, then this sentence fails to represent the world accurately, and we can say that it is false.

Moral language has an intriguing mixture of features which seems to straddle the representational and the non-representational. A sentence such as 'You ought to put the cat on the mat' is a sentence which can be aptly described as true or false. But it is hard to identify what sort of fact is being represented by such a sentence. And if there is a fact that corresponds to this sentence, then it is a particularly odd sort of fact: it is one that is supposed to *motivate* us to act in a particular way.

Consequently, a number of proposals have been made to understand normative language in ways which do not require that normative language gets its meaning by representing facts, yet can still explain the grammatical features of that language which make it resemble normal representational language. This broad programme is known as *non-cognitivism* about ethics. Ethical non-cognitivists are committed to denying that moral sentences function by representing facts. This idea is sometimes put a bit differently, as the idea that moral sentences do not have substantive truth conditions.[12] Non-cognitivists usually also make a psychological claim: that the psychological state which we *express* through our moral utterances is not belief-like. It is a conative rather than a cognitive state.[13]

Non-cognitivism has received extensive attention as a proposal for normative language such as moral and evaluative language. It has been much less closely examined as a proposal for chance.[14] The question now facing us is: is non-cognitivism a better approach than the other forms of anti-realism?

---

12 It is necessary to qualify the truth conditions as 'substantive', because the non-cognitivist does not wish to deny that moral sentences may have *trivial* truth conditions, such as: 'It is true that "*p*" if and only if *p*.'

13 For a more thorough discussion of the distinction between cognitivism and non-cognitivism, which also helpfully reviews many of the main arguments in this area, see van Roojen (2011).

14 The only recent exceptions of which I am aware are Logue (1995) and Ward (2005). Both Logue and Ward propose an understanding of probability claims (not just chance claims) in quasi-realist terms, allying it to the quasi-realism of Simon Blackburn (1984; 1993). Note that Briggs (2009b) has argued that Ward's account, at least, still encounters fatal difficulties due to his underlying Humean metaphysic, which leads to the 'big, bad, bug' – an objection to David Lewis's theory of chance.

The viability of non-cognitivism is a difficult question to judge, in large part because the doctrine is most easily characterised by what it does *not* say than by anything it does say. That said, non-cognitivism faces a very serious challenge, which can be framed in terms of the following question: *is all language non-cognitivist?*

Call the person who answers 'Yes' to this question a *global* non-cognitivist (e.g. Barker 2007). The global non-cognitivist is committed to a wholesale revision of our philosophy of language. Tools and ideas that have become deeply entrenched in philosophical logic, such as possible-worlds semantics, will be largely or entirely obsolete. Instead, something like the Gricean framework for analysing meaning in terms of speech acts must come to the fore.

If we adopt a global form of non-cognitivism, then we had better come up with a very good account of non-cognitivist semantics. This is hard work, and well beyond my ability to discuss rigorously here.

But also note that if we adopt a global form of non-cognitivism, we appear to lose one of the superficial attractions of adopting non-cognitivism about chance. That is, if chance facts appeared to be metaphysically mysterious, in comparison with non-chancy facts, then there would be an apparent benefit in claiming that chance sentences are non-factual, compared to non-chance sentences. The two sorts of sentence stand in very different relations to the world, so there is no need to theorise about mysterious chance facts in order to make sense of chance sentences. If, however, even the sentences that we think of as *unproblematic* work by some non-cognitivist means, then we have seemingly thrown out the baby with the bathwater. We no longer have a division of linguistic function to match the division between metaphysically easy and metaphysically difficult cases.

If we answer 'No' to the above question, thereby adopting a form of *local* non-cognitivism, then we face even worse problems. As Mark Schroeder (2008) has argued, it is very difficult to develop a systematic semantic theory that mixes non-cognitivist and cognitivist sentential components. And yet we mix chance claims in with non-chance claims all the time.

So there is a lot of work to be done by the non-cognitivist if the programme is to succeed. But I don't want to give the appearance that we are yet in a position to adjudicate on this matter on grounds of semantics. As I have indicated, I take subjectivism to be an irretrievable position. But both error theory and non-cognitivism appear to be viable alternatives. There might be some degree of indeterminacy in our linguistic practices as to whether they are best interpreted as objective, but in error, or as non-cognitive. And it will require much more detailed work in philosophical semantics to determine whether one approach or the other can decisively get the upper hand.

It is more productive to focus on the very first problem noted with anti-realism about chance: the apparent implication that it threatens scientific realism. To address this concern, I propose that we return to a closer examination of the way chance features in what has been described as the most successful empirical theory yet produced: quantum mechanics.

In trying to develop a theory of chance, I have so far drawn largely on a metaphysical picture inspired by classical physics, and have used statistical mechanics as my central example of chance arising in physical theories. It has proven difficult to develop an adequate theory of chance. But perhaps things will look different if we turn to quantum mechanics. The theory of quantum mechanics is extremely well confirmed. It is doubtful whether it is strictly true in its current form, due to the well-known conflict between quantum mechanics and the general theory of relativity, but it is an outstandingly good example of a successful physical theory, and it represents a stark break from earlier models of the physical world. Perhaps by tapping into these resources, we can develop an adequate metaphysics of chance.

39 The quantum mechanical world

It is widely believed that quantum mechanics is starkly opposed to classical physics, because quantum mechanics claims that the world is governed by fundamentally indeterministic laws. As it happens, this common belief oversimplifies somewhat. Quantum mechanics is a theory that is formulated in relatively mathematical terms, quite removed from concepts of directly observable physical entities. Consequently, there is a great deal of room for interpretation of the meaning of the mathematics. Indeed, there are at least three interpretations of quantum mechanics which are serious candidates for giving an adequate account of how the mathematics relates to reality.

Of these three interpretations, one of them – Bohmian mechanics – is completely deterministic. The other two interpretations each involve probability, but in rather different ways. So there is no straightforward answer to the question, 'What is the role of probability in quantum mechanics?'

Determining the correct interpretation of quantum mechanics is utterly beyond my ability. All that I can hope to do is to throw some light on the role of probability in quantum mechanics, on each of the three interpretations. It will turn out that two of the interpretations are relatively 'conservative' of our prior understanding of probability: they will suggest no new model

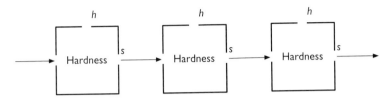

Figure 9.1 An electron, measured as soft, will continue to be found soft on subsequent measurements.

for how chance might arise. But the third – the so-called Everett or 'many-worlds' interpretation – suggests an altogether novel metaphysics of chance.

40 Weird quantum phenomena

Quantum mechanics predicts the behaviour of objects such as electrons. The properties of electrons include familiar ones, such as electric charge, position, and mass. But there are also other properties, studied by quantum mechanics, which are relatively unfamiliar. The details of these unfamiliar properties need not bother us, so following David Albert, I will give them some familiar names.[1] Electrons have properties that I will call *hardness* and *colour*. An electron's hardness involves it being either hard or soft. An electron's colour involves it being either black or white.

These properties of electrons are measured or observed by firing the electrons into devices that we will call *boxes*. An electron fired into a hardness box will come out of one aperture if it is hard and out of the other aperture if it is soft. Similarly for a colour box. So measurements involve separating electrons with different hardness and colour properties into different *physical positions*.

In one sense, hardness and colour properties seem to be quite stable, in that, if you measure the hardness of an electron once, and it comes out of the soft aperture, it will continue to be found soft on subsequent measurements (see Figure 9.1). The same goes for hard electrons, white electrons, and black electrons.

But there is another way in which these properties are seemingly unstable, and it suggests that hardness and colour are interrelated. If you take a bunch of black electrons, and measure their hardness, each one is 50 per cent likely to be hard or soft. (How do we know these electrons are black? Because

---

1 The account I give here is heavily indebted to Albert's extremely lucid presentation, in *Quantum Mechanics and Experience* (1992).

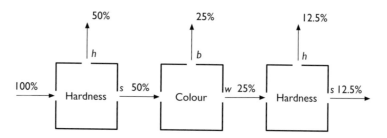

Figure 9.2 A series of measurements of hardness and colour. The percentage figures show the proportion of electrons from a typical sample that will emerge from each aperture. Note that if hardness and colour were ordinary properties, you would expect all the electrons that enter the second hardness box to emerge from the soft aperture.

we have just seen them emerge from the black aperture of a colour box.) Similarly, if you take a bunch of white electrons, they will be 50 per cent likely to be hard, 50 per cent likely to be soft.

This probabilistic phenomenon seems to be *irreducible*. We can find no property of the black (or the white) electrons that correlates with whether or not they are going to be hard or soft. (And the same thing happens with hardness. If you measure hard electrons – or soft electrons – for colour, then they will each be 50 per cent likely to be black, 50 per cent likely to be white.)

Now, suppose we start out with some black electrons, and then measure them for hardness. As noted, roughly half of them will be hard and half of them will be soft. If we take just the hard ones, and measure them for colour, what would we expect? If these were classical properties, you'd expect all the electrons *still* to be black. All we have done, by measuring them, is sort them out into black-and-hard versus black-and-soft.

But that is not what we see. Instead, we find that each electron has a 50 per cent probability of being black and a 50 per cent probability of being white. The same thing happens if we take the soft electrons, which were originally black. If we measure them again they will now be, on average, 50–50 black and white. And the same thing happens if you take some white electrons and measure their hardness. They end up 50–50 black and white. Measuring them for one pair of properties seemingly 'scrambles' the electrons for the other pair of properties. See Figure 9.2 for a typical series of measurements.

Because of this apparent scrambling, for all we have been able to discover about electrons and colour and hardness, it appears to be *physically impossible* to measure them for both of these properties. This is a bit odd, but we can live with it: these properties are obviously a bit different from ordinary hardness and colour properties. The properties are instead *interdependent*

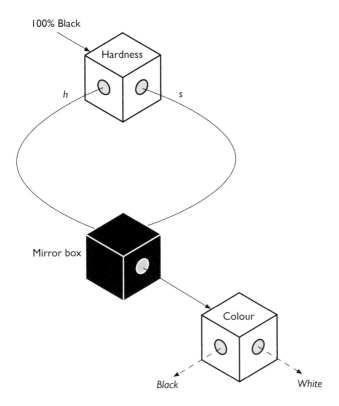

100% Black

Hardness

h

s

Mirror box

Colour

Black

White

Figure 9.3 All electrons fed into the hardness box are black. Each electron then has a 50 per cent chance of taking the path *h* or the path *s*. After completing that journey and going through the mirror box, a colour measurement is made. What do you expect the proportion of black and white electrons to be?

in an interesting way. That may be unfamiliar, but electrons are unfamiliar things, so it is not too alarming to find that they have somewhat unfamiliar properties.

But now consider a much stranger thing. Suppose I create an apparatus as in Figure 9.3. One at a time, I feed black electrons into a hardness box. Coming out of that box, there is a path that the hard electrons take (*h*) and a path that the soft electrons take (*s*). The two possible paths are configured so that they converge on a second box – call it a 'mirror box' – that takes soft electrons in one aperture and hard electrons in the other, and puts them both out via the same aperture. The mirror box does not change the properties of the electrons that pass through it in any way.

After the mirror box, we put a colour box.

Now, feed some black electrons into this apparatus. We expect each electron we feed in to have a 0.5 chance of emerging from the hard aperture, and a 0.5 chance of emerging from the soft aperture. We also expect, because of the way these measurements scramble colour, that half of the electrons coming along each path are black, and half are white. (Indeed, if we put a colour box on the paths, that is exactly what we see.)

Then, regardless of whether the electron was hard or soft, it will enter the mirror box, and emerge from the single aperture. It will then enter the colour box. What will we get? Surely, because these electrons have all been measured for hardness, we should expect them to manifest the characteristic scrambling: 50 per cent chance of black, 50 per cent chance of white.

But that is not what we see! All of the electrons that enter the colour box are black.

This is very strange. How do the electrons change back? We have never been able to separate hard and soft electrons without scrambling their colour. Now, having created two paths which separate the hard from the soft, and then arranging things so that those paths reconverge, the result is that the electrons behave as though the measurement has never been performed.

What happens next, if you place a barrier in one of the paths, as in Figure 9.4? Now, you might figure, we should get only half as many electrons arriving at the colour box, but they should all still be black.

Well, you are half correct. We do get only half as many electrons arriving at the mirror box, but now half are white, half are black!

This is extremely strange. Consider a single soft electron that takes path *s* in the *first* case, without the wall. When it arrives at the mirror box, it will definitely, always, no matter what, come out black. Now consider an *intrinsically identical* soft electron that takes path *s* in the *second* case. The only thing different, as it takes its journey along *s* to the colour box, is that path *h* is now blocked. So, seemingly, *no difference is made on the path that the electron actually takes*. The difference exists merely at a remote location, on a path that the electron *could have taken*. But this difference is enough to make it the case that when the electron arrives at the colour box, it is now a matter of chance whether or not the electron is black.

It becomes clear from considering cases like these that in quantum phenomena we are dealing not *merely* with irreducible probabilities. Those are hard enough to take. We are also dealing with very strange phenomena, which seem to involve *non-local interactions*. Whether or not there is an obstacle on a remote path which an electron might have, but did not travel down, makes a difference to how the electron behaves when we later measure it.

How does the theory of quantum mechanics explain this?

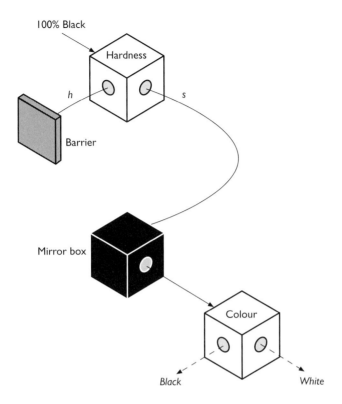

100% Black

Hardness

h   *s*

Barrier

Mirror box

Colour

Black   *White*

Figure 9.4 The same apparatus as before, except that now path *h* is blocked. Roughly half of the electrons will emerge from the soft aperture, but strike the impeding wall, and not make it to the mirror box. In this condition, what proportion of electrons measured for colour at the end will be black?

41 The formalism of quantum mechanics

Quantum mechanics describes the world at a particular time by means of a mathematical object: a state vector (or 'wave function'). The state vector evolves deterministically, according to a dynamical law described by the Schrödinger equation.

There are two sorts of state into which the state vector can evolve over time. Sometimes – in the happy cases – it evolves into states which we can unproblematically associate with an observation or a measurement. We might write such a state |Up⟩ and that simply means that a given particle has spin up (or you can call this property 'hardness' if you prefer). If you prepare a quantum system containing a particle such that it evolves into this state and observe its spin state, you should see that particle with spin up.

Sometimes, however, the state vector evolves into less happy cases, that appear – to a naive eye – to be 'weighted sums' of the happy states. These are the states known as 'superpositions'. An electron, for instance, could be in a superposition of a spin-up state and a spin-down state. That would be written something like:

$$\frac{1}{\sqrt{2}}|\text{Up}\rangle + \frac{1}{\sqrt{2}}|\text{Down}\rangle$$

The '$\frac{1}{\sqrt{2}}$' terms are the amplitudes of the so-called 'component vectors' of this wave function. There is a rule governing these terms, such that, when they are all squared, they must sum to one. Subject to that constraint, they can take any value, including negative ones.

Superpositions are the source of much of the puzzlement over what quantum mechanics means for the nature of the world. One of the most famous thought experiments involving a superposition is the case of Schrödinger's cat. This is the case of a cat which is placed in an apparatus which has the potential to poison the cat. The release of the poison, however, is tied to a quantum mechanical device, such that the apparatus evolves into a superposition involving one state where the poison is released and the cat is dead, and one state where the poison is not released and the cat is alive. The key point of the example – for our purposes – is to emphasise that the strangeness of superpositions is not a phenomenon that is restricted to subatomic particles. Nor is it a phenomenon which only applies to unfamiliar properties such as whether or not something has spin up or down. Quantum superpositions, rather, can involve everyday macroscopic objects such as cats, and can involve matters as serious as life or death.

The superposition involved here could be written something like:

$$\frac{1}{\sqrt{2}}|\text{Alive}\rangle + \frac{1}{\sqrt{2}}|\text{Dead}\rangle$$

What does it mean, though, for the world to be in such a state?

It is best to approach this question by focusing on the role of superpositions in quantum theory. Superpositions do two things for us. First, and most straightforwardly, they give rise to predictions that we will see probabilistic phenomena. In particular, the *amplitudes* of the component vectors determine the probability that we will see one or the other component vector of a superposition. In the case of Schrödinger's cat, if we prepare a superposition of alive and dead, this will lead to us seeing either a live cat or a dead cat. If we repeat the experiment, we will see statistical frequencies of live cats and dead cats which support some probabilistic hypotheses more strongly than others. And we can perform different experiments, in which the amplitudes *a* and *b* have different values, such that we observe different frequencies in the statistics. There is a simple mathematical rule, called the Born rule, which we can use to turn the amplitudes into probabilities, and it has been extremely well confirmed: the statistics match the predictions of the Born rule extremely well.

Second, the mathematical properties of superpositions, as governed by the Schrödinger equation, are such that they can be used to predict the very strange sorts of interference and non-local phenomena that have been observed in quantum mechanical experiments, such as the two-slit experiment, or the experiment with the two paths, described above.

So superpositions are *required* to explain the process by which the world evolves and gives rise to strange, non-classical effects. *But*, we seemingly never observe superpositions. Every time we look, we see a normal state that is definitely one way or the other. We see a cat that is alive, or a cat that is dead.

This does not suffice to answer our original question: *what are superpositions?* – but it helps to give us some context for why the theory has them at all. In order to come to grips with some concrete proposals for what superpositions are, we need to focus more closely on the sorts of chances we see in quantum mechanics.

42 Chance in quantum mechanics

As noted above, there is a simple rule – the Born rule – which allows us to derive from the mathematical representation of a superposition what the probability is of observing a given component of that superposition. And these probabilistic predictions are extremely well confirmed.

Where did this probabilistic phenomenon arise from, given the state vector deterministically evolved into a superposition? How does the superposition give rise to this probabilistic phenomenon?

It is worth stressing the oddness of the partial story we have told so far. (1) The wave function evolves in accordance with a strictly *deterministic* dynamical equation: the Schrödinger equation. (2) The states that the world evolves into, according to the Schrödinger equation, are such that they inevitably involve mysterious superpositions. But (3) we have never knowingly observed a state that is a superposition. We observe electrons with spin up or with spin down. We observe live cats or dead cats. We

Figure 9.5 The two processes by which the world evolves, on a collapse interpretation.

never find ourselves in a state in which an electron, or a cat, appears to be 50 per cent in one state and 50 per cent in the other.

Moreover, if we attempt to create an experimental circumstance in which – with certainty – a superposition like the following will occur,

$$\frac{1}{\sqrt{2}}|\text{Up}\rangle + \frac{1}{\sqrt{2}}|\text{Down}\rangle$$

then we take ourselves to have *succeeded* if we find that about half the electrons coming out of the device have spin up, and about half have spin down. That is, we expect the amplitudes of the states in a superposition to correspond – in some probabilistic fashion – with the observed frequencies of those states in our observations.

Explaining this role of the amplitudes is an important constraint on any adequate interpretation of quantum mechanics. But it also requires us to think about the nature of chance itself, since these amplitudes seem to determine objective probabilities.

Some interpretations of quantum mechanics postulate a second, distinct type of process by which the world evolves. This is the notorious idea of 'collapse'. The thought is that sometimes, instead of evolving in accordance with the Schrödinger equation, the world evolves by 'collapsing' into a state that is not a superposition, but is a perfectly determinate state. And the probability that a superposition collapses into any one of its component states is determined by the amplitude of that component.

The alleged collapse of the wave function is at the heart of controversies over the interpretation of quantum mechanics. The problem for a collapse interpretation is roughly this: a collapse view entails that there are two incompatible ways for the wave function to evolve. It may evolve in the strictly deterministic way described by the Schrödinger equation, or it may evolve in the strictly stochastic, chancy way, known as collapse. *What makes it the case that the world evolves in one of these ways, rather than in the other?* This is the crucial question that any collapse interpretation of quantum mechanics must address. (See Figure 9.5.)

The Copenhagen interpretation of quantum mechanics notoriously advocated the idea that it was a question of whether or not the system in question was being *observed*. This gives rise to very serious difficulties, because 'observation' is not a well-defined physical term, and every attempt to give it a precise physical meaning leads to some absurdity.

The three broad alternative schools of thought that have currency[2] can be summarised as follows:

An independent law for collapse

Some accounts suggest that collapse is an independent physical process that happens to the wave function in accordance with some separate law. The best known of these proposals, due to Ghirardi, Rimini, and Weber (GRW), claims that the additional process is an irreducibly stochastic one: for any amount of time, there is a determinate probability that the wave function will spontaneously 'localise' in that time. That is, the shape of the wave function will change in such a way that it ceases to be in a superposition, but instead is in one of the 'happy' states that can be uncontroversially associated with an observation.[3]

Such an account does not say that superpositions cannot happen. We can certainly find ourselves in them. But the posited law of spontaneous collapse entails that any superpositions we are involved in are extremely unlikely to last very long. In part, this is because the probability of collapse depends upon the number of interacting particles involved in a system. As the number of particles increases, the chance that one of them will induce a collapse increases greatly. Consequently, we should never expect to find ourselves – being constituted of very large numbers of interacting particles – in a superposition.

So this approach answers the crucial question above by saying the world is *simultaneously* subject to both laws. The Schrödinger evolution is the default, but when the stochastic law 'fires', it 'overrides' the Schrödinger evolution.

Hidden variables

Some theories postulate additional ontology in the world to explain what looks like collapse. The most prominent example is Bohmian mechanics,

---

2 Maudlin (1995) argues that, if framed more precisely, these are the only three logically possible categories of response to the measurement problem.

3 Ghirardi, Rimini, and Weber (1986). See also Albert (1992: 92–111) for a sympathetic, though critical, discussion.

which denies the assumption – which features in at least some versions of spontaneous collapse theories – that the state of the world is *exhausted* by the wave function. Instead, on Bohm's account, the wave function is conceived of as something like a field that influences the trajectory of all the particles in the world. The particles are all possessed of entirely determinate, ordinary properties. But – here's the 'hidden' bit – the exact positions are not known: they are physically unobservable.[4]

On this view, the evolution of the world is entirely deterministic. The apparent chanciness that arises in quantum mechanical experiments is due to our inability to know the initial conditions of our experiments. If we could measure the positions of the particles more precisely, there would be no probability in quantum mechanics. But because there are fundamental limits on our ability to measure the positions of the particles, we will never be able to eliminate the apparent chanciness.

No-collapse accounts

A no-collapse approach, as developed by Hugh Everett, simply denies that collapse ever occurs![5] This is, in many ways, the most startling of all the interpretations, because it seems to fly in the face of our experience. As indicated above, we do not seem to *see* superpositions. But the Everett interpretation implies that superpositions are ubiquitous.

Supposing this problem can be overcome, however, the Everett interpretation has the advantage of theoretical parsimony over the other interpretations. It advocates the very same wave function that all the interpretations need, but dispenses with the need for any additional ontological posits – as in a hidden variable account – or any additional processes of evolution – as in a spontaneous collapse account.

What might these theories, if true, show us about the nature of chance? In particular, can any of them suggest a better understanding of how there could be chances that meet the various constraints laid down in the opening chapter?

---

4 Bohm's theory is contained in Bohm (1952). An influential presentation of the idea is given by John Bell (1982). There are other hidden variable accounts which are not deterministic, but I'll set them aside here, as they seem to have nothing new to tell us about chance.
5 The central references are in DeWitt and Graham (1973); Everett (1957a, b). For a more philosophical examination of the idea, see Barrett (1999).

What we learn about chance from spontaneous collapse

A spontaneous collapse account such as the GRW theory explains the fact that we see only a live cat or a dead cat with a certain probability by appeal to a dynamical law that governs the collapse of the wave function. That dynamical law is itself probabilistic. So one probability – the chance that we see the cat alive or dead – is explained by another – the chance that the wave function will collapse (or 'localise') in a given period of time.

This account is compatible with various of the philosophical accounts of chance that we have examined. We could think that the chances are underwritten by complicated facts about the actual world, as the Humean claims. Or we could think that the chances are in some way primitive, and can only be explained in terms of possibilities if we help ourselves to a circular appeal to chance-like notions.

In any case, I don't think there is anything much that we can *learn* about chance from the spontaneous collapse proposal, because such proposals rely directly and explicitly on a claim about *chancy* collapse. Given that chance is the very thing that we are attempting to understand better, we need an account that explains chances in terms of something else. So for metaphysical purposes, we can move on.[6]

What we learn about chance from hidden variable accounts

Hidden variable accounts, such as Bohmian mechanics, make chance in some sense epistemic. Because we cannot know the precise positions of the particles, we cannot predict with certainty what will happen to those particles in future. So we start with a subjective probability distribution over initial states, and we use our knowledge of the dynamical laws to determine an appropriate distribution over final states.

Put this way, the probabilities we end up with do not look much like chances, but rather like subjective probabilities. So one might simply draw the sceptical conclusion from a deterministic theory like Bohmian mechanics that there are no chances. But as we saw in the case of the classical

---

6 Roman Frigg and Carl Hoefer (2007) claim that the probabilities in the GRW interpretation need to be interpreted either in neo-Humean terms, along the lines of David Lewis's account, or as propensities. For conceptual reasons, they prefer the neo-Humean alternative. Apart from their hasty dismissal of anti-realist accounts of chance, I am inclined to agree. Indeed, I take their argument to weakly confirm my view that we get little new purchase on the nature of chance from spontaneous collapse interpretations of quantum mechanics.

picture, it might still be possible to defend the idea that there are credal states – other than outright certainty – which are objectively *best* to have, given the available evidence. If that is so, then there is still room to identify chances in such a picture. One way this might happen is if the appropriate initial probability distribution simply held its status as best as a matter of objective fact. This would give us something like a modal volume theory of chance for Bohmian mechanics, analogous to the theories discussed in Chapter 6.

Another, more subtle way for something chance-like to be explained in terms of this picture is by using the idea of *microconstancy* described in §29. The idea would be that, given the nature of the dynamical laws, for any initial subjective probability distribution you begin with (provided it is sufficiently smooth), given the microconstancy of the dynamics, it is reasonable to ascribe probabilities to the outcomes of a quantum experiment that conform very closely to the Born rule. So we have objective reason to think that, even if we do not know much about the initial conditions of the system, the likely outcome over a large collection of trials is that the frequencies of different behaviours will match the frequencies predicted by the Born rule. Maudlin (2007b: 290) claims that this is the correct understanding of the probabilities in Bohmian mechanics. However, he denies that these are chances for individual events: 'they are rather typical empirical distributions for large sets of events'. Strevens (2003: § 5.6) takes a much bolder view. He suggests – though with considerable caution – that the microconstancy account of probability gives us some reason to be suspicious that there are any fundamental and irreducible chances. Hence, a deterministic interpretation of quantum mechanics is supported, not on standard empirical grounds, but on grounds of its philosophically superior account of probabilities!

As I argued in §29, however, although the microconstancy account is a clear improvement over standard possibilist accounts of chance, it remains committed to the objective 'preferability' of some measures over the space of possibilities rather than others. As such, it gives rise to many of the same concerns as possibilist accounts. What makes these measures right? How do we learn the correctness or otherwise of a measure? What role can these measures play in causal explanation? Instead, I suggest that Strevens's enterprise supports the anti-realist enterprise by showing a source in the world for the *apparent* objectivity of chances: that is, even agents with wildly divergent initial prior opinions will, given sufficient knowledge of the dynamical laws, converge upon very similar credences for outcomes.

What we learn about chance from no-collapse accounts

The Everett interpretation of quantum mechanics (EQM) is in many ways the most elegant response to the problem of collapse. It does not posit any additional ontology. Nor does it posit any additional theory to explain why and when collapse happens. Rather, it *denies* that collapse happens. Consequently, the Everettian claims that, in some sense, *all the component states of a superposition happen.*

How can that be, given that the component states of a superposition are apparently incompatible: either the cat is alive or it is dead? It surely cannot be both. Here, subsequent defenders of the Everett idea – most notably Bryce DeWitt – have employed a metaphor of branching.[7] Whenever a superposition occurs, then in some sense the world develops separate branches. In a branching event – what we would ordinarily think of as a chance event – *all* possible outcomes of a chance phenomenon occur, but they occur in discrete space-times that do not – at least as a general rule – interact in future.

To take a familiar macroscopic example: assume that if a coin is tossed it is chancy whether or not it will come to land heads or tails. On a branching world metaphysic, this means that *both* events happen. But each happens in a region of the universe that is, for all intents and purposes, physically isolated from the other. And all other events fall in line with the branch-event. In one branch, observers will see the coin land heads up. In the other branch, observers will see it land tails. And so on and so forth.

As we have already noted above, superpositions are the mathematical device by which quantum mechanics gets probabilistic phenomena. So this metaphor of branching is really a metaphor that is to be applied to superpositions. Suppose you put Schrödinger's cat in the box. A superposed state develops in which there are two component states: a live-cat state and a dead-cat state. According to the Everettian, even after we open the box, both states – in some sense – exist. There is a future branch of the cat which is

---

7 See DeWitt and Graham's preface to DeWitt and Graham (1973). See also discussion in Barrett (1999: chap. 6) and Wallace (forthcoming: chap. 3).

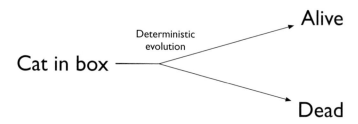

Figure 9.6 Schrödinger's cat, on the Everett interpretation.

alive, and there is a future branch of the cat which is dead (see Figure 9.6). *We do not observe an amalgam of cats, because we too enter a superposed state. So consequently, we too have branched.* That is: someone looking into the box will go into a superposition of (i) seeing a dead cat and (ii) seeing a live cat. That means the observer will branch into one person who accompanies the dead cat and another who accompanies the live cat. Moreover, there is a story that can be told to explain – claim the Everettians – why the branches cannot interact, except in very special circumstances. The sorts of circumstances in which we see branches interact are the circumstances in which we see bizarre non-local phenomena, such as those described in §40.[8] At the larger scales where we usually act and intervene – at the level of cats – such interference does not happen for reasons explicable by reference to the original quantum mechanical formalism.[9]

　　If we try to consider the idea of branching as a physically real matter, it is admittedly extremely puzzling.[10] How can one space-time spawn multiple others? It sounds as though such a process would violate all sorts of other physical constraints such as conservation laws. But for these reasons it is

---

8　David Deutsch forcefully presents the Everett interpretation, in chapter 2 of *The Fabric of Reality* (1997), as the most straightforward and natural interpretation of quantum interference phenomena.

9　Everettians explain this by appeal to *decoherence* of the wave function. It is beyond my expertise to evaluate the success of this part of the Everettian programme, so I shall simply grant it as an assumption. See Wallace (2003) and references therein for discussion.

10　Note, the Everettian idea of branching is importantly different from accounts which ontologically privilege one branch over the others, such as Storrs McCall's (1994). McCall thinks of the evolution of the universe as involving something like a process of branches falling off a tree. Whenever a moment becomes *present*, only one branch is 'selected'. Other branches that are distinct from the selected branch at that time cease to be real. Thus McCall inherits many of the problems that affect collapse accounts. What explains the fact that the world takes one branch rather than others?

　　On the sorts of branching accounts I am discussing, all branches are real: we are on just one of many. And there are other branches that share a common ancestry with us: our causal antecedents have spawned numerous causal offspring.

best to stress that the idea of branching is to some extent metaphorical. And there are some clear limits to the many-worlds or world-branching metaphor. Here are two: there is no determinate time at which two parts of the state vector branch; nor is there a determinate number of branches at any given time.[11] In some physically explicable sense, *there are branches* where the cat is alive, and *there are branches* where the cat is dead. But there is no physically privileged *number* of branches that correspond to these descriptions. There is, however, a physical quantity, called the amplitude, associated with the component vectors of a mixed state. And this is the quantity which the Born rule tells you how to turn into a probability.

What the Everettian says about the reality of superpositions is in some ways congenial. We need those superpositions in order to explain the strange quantum phenomena I described above, and the Everettian account makes the apparent non-locality of quantum phenomena turn out to be a local phenomenon in the higher-dimensional space in which the many world branches exist. But Everettian quantum mechanics makes it hard to explain how the Born rule works to give rise to precise probabilities. In reality, the Everettians say, *every possible outcome* from a probabilistic experiment *happens*. Yet we observe only a fraction of the possible outcomes for each experiment. Moreover, the outcomes we observe occur in frequencies that correspond with the Born rule. The Born rule tells us how to convert amplitudes of component vectors into probabilities that we will observe the corresponding state. But if we conceive of component vectors as equally real branches, there is no physical story that has been told about why the amplitudes give rise to the probabilistic phenomena that we see. Having been told that *everything* will happen, why should I ascribe any particular probability to the claim that I will see one particular possible outcome? This topic warrants extended discussion.

---

11 The case for indeterminacy about the details of branching is made in Wallace (2007: 328–9). Though see P. Lewis (2010) for a contrary view on this.

As we saw in the previous chapter, the Everett interpretation seems to dispense with probabilities in quantum mechanics. Instead of describing a world for which many things are possible, but only some of those possibilities are actualised, it suggests a world in which *all possibilities actually happen.* Moreover, this is something of which we can be absolutely *certain.*

Obviously, this does not straightforwardly fit our experience. We do not see multiple possibilities becoming actual. Whenever we measure a superposed particle, we observe only one property or another. Moreover, we have a very useful probabilistic rule to help us to predict what we will see. How can the success of this probabilistic rule be explained, if an Everett world does not involve any uncertainty?

There are two main moves that are employed in response to this challenge. The first – what I call Stage A – is to show that there is a relevant sort of uncertainty, even in a universe where we are certain that everything will happen. The second move – Stage B – is to try to vindicate the probability of the Born rule in particular. That is, to show that we should not merely be uncertain about the future, but that we have good reason to attach the particular probabilities dictated by the Born rule to the possible outcomes of a quantum experiment. We will consider these two stages in turn.

43 Uncertainty in an Everett universe

Everettians encourage thinking about branching as akin to a process of splitting, or 'fission', that is much discussed in the literature on personal identity.[1] Here is an example of the genre, adapted from Derek Parfit (1985).

The two ships (background) Our subject, DP, is a mechanic with expertise in repairing spaceships. It is very costly to have the spaceships land, so they usually orbit the Earth, and he arranges to be teletransported to the ship he will work on. The

1 See, e.g. Saunders (1998); Wallace (2002, 2007).

teletransportation works by accurately measuring the location and nature of all the particles in the traveller's body. The traveller's body is then destroyed, and a perfect replica made at the destination, using the information that has been recorded at the origin.

The first question to ask is: is this really a means of *travelling*? If DP uses this method, does he survive the teletransportation process, and successfully travel to the remote spaceship, or has he in fact caused his own demise, and merely created a very good replica of his former self on the spaceship?

For our philosophical convenience, we will stipulate that the traveller does indeed survive the process of travel. This is a controversial thing to stipulate: many philosophers think it presupposes a dubious view of personal identity.[2] But we could readily modify the mechanism by which the transportation works so as to avoid this concern. The purpose of this story is merely to provide an analogy with what happens when the world branches in an Everett universe, and the doubts some people have about survival through teletransportation do not apply to survival through world-branching. So suppose, for the sake of the argument, that you subscribe to an account of survival which fits this mechanism.

Now, on with the story:

The two ships (continued) In this instance, DP intended to go to the spaceship *Potemkin*, and not to the *Enterprise*, which is a ship he had worked on the previous day. Before activating the teletransporter, DP says to his co-worker: 'I'm going to *Potemkin*.'

Today, however, the device malfunctions. Instead of creating only one replica of DP's body on *Potemkin*, a *second replica* of his body is created on *Enterprise*. The original body on Earth is destroyed as usual.

Ordinarily, had only one replica been created of DP's body, we would have confidently said that DP survived in that new body. But here it is not clear whether and how DP has survived. There seem to be some grounds for saying that DP has survived, both by travelling to *Potemkin* and by travelling to *Enterprise*. But in that case, how can one person have become two? Exactly what to say is unclear.[3]

---

2 Bernard Williams, for instance, puts forward an opposing view about personal identity towards the end of his 'The Self and The Future' (1970). This sort of view requires bodily continuity for survival. As will become evident, the analogue of a fission case in an Everett universe would indeed involve bodily continuity, so it should address Williams-type concerns.

3 I expect some readers will find these scenarios infuriatingly artificial and unrealistic. If the Everett interpretation is correct, however, something importantly similar to this process is

Imagine, though, that *you* are DP, and you have just undergone this experience of malfunctioning teletransportation. There is a reasonably widespread and robust intuitive response to such stories, to the effect that: '*I* will become one or the other fission product, but not both.' It is possible to imagine going through this process and finding oneself on *Potemkin*. It is also possible to imagine going through this process and finding oneself on *Enterprise*. But it is seemingly not possible to imagine being both on *Potemkin* and on *Enterprise* at the same time.

So we seem to be committed to the thought that there are *two possible outcomes* here:

1. I survive by arriving on *Enterprise*.
2. I survive by arriving on *Potemkin*.

Moreover, I seem to be legitimately *uncertain* as to which of these two outcomes will occur.

If this thought is correct, then perhaps it can be applied to the sort of branching that occurs in an Everett universe. For instance, think back to the superposition that arises in the case of Schrödinger's cat. Having opened the box, the world has evolved into a superposition of two states:

• You observe that the cat has died.
• You observe that the cat has survived.

If this can be understood in terms of the metaphor of branching, then something very like the story of the two ships has occurred, but it has happened on the scale of the entire universe. Not only have you undergone a fission process, but the entire world has branched, such that there is one world-like entity in which there is a replica of your former self observing a deceased cat; and there is a second world-like entity in which a replica of your former self is observing a live cat.[4] Corresponding to these two fission products, we can imagine two possible outcomes:

1. I survive by arriving on the dead-cat branch.
2. I survive by arriving on the live-cat branch.

---

happening to us all the time! So it will pay to be somewhat tolerant and to attempt to give this idea a fair hearing.

4 Note that branching is a process that does not happen everywhere instantaneously. So by saying 'the world has branched', I do not mean to imply that this process happens all at once, thereby defying special relativistic prohibitions on faster-than-light processes.

Moreover, there is genuine uncertainty as to which of these states will be actual. So it appears that, if the intuitive response to fission cases is legitimate, the Everett view will allow that there can be genuine uncertainty after all.

But the intuitive response to fission cases is widely regarded with a great deal of philosophical suspicion, because it appears to be underwritten by a 'non-reductionist' view of personal identity. That is, it suggests that, over and above the facts of physical continuity, psychological connectedness, and the like, there is a *further fact* about personal identity. Our intuition about a fission case appears to be a manifestation of the belief that this further fact about identity can turn out one way or the other, despite holding *all the other facts fixed*.

Here is a way of spelling out this concern into something like an argument.

P1. Whenever we correctly imagine that there are two distinct possibilities, then there must be two distinct ways for the world to be which correspond to those possibilities.

P2. In superpositions (or other fission cases), we correctly imagine that there are two distinct possibilities: (i) observation of one outcome and (ii) observation of the other.

Therefore,

C1. There must be two distinct ways for the world to be which correspond to the possibilities of (i) observing one outcome and (ii) observing the other, arising from a superposition.

But this contradicts our understanding of an Everett universe. The Everett universe is supposed to be *completely* described by the state vector. According to that representation, there is just one way for the world to turn out in a given superposition. There is no further fact that determines what it is that is observed. Similarly, in science fiction cases such as the two ships, it is argued that there are not two distinct ways for the world to be which correspond to the two imagined possibilities. There is just one way for the world to be, completely described by the physical facts involving dual teletransportation. To claim that there are two distinct possibilities seems to invoke a mysterious non-physical fact about personal identity – a fact that obtains in one possibility and not in the other. But such a non-physical fact contradicts our working assumption that physics gives a complete description of the world.[5] Worse, it is not clear that it will account for our ideas about survival without creating new difficulties. (How would

---

5 This point is made very clearly by Dilip Ninan (2009: §3). See also Parfit (1985: chaps. 10–11).

we ever be able to know facts about personal identity, if they float free of all physical facts? How do these identity-making facts causally impact other events? And so on.)

How should the Everettian respond to this apparent clash between our intuitive response to fission cases and our understanding of the world as exhaustively described by the state vector? There are roughly two responses, each rejecting one or other of the two premises above.

The typical response has been to reject (P2). That is, to reject the intuition that there are two or more possibilities ahead of us, as we enter a superposition. There is only one possibility, but it is a possibility in which we undergo the peculiar experience of branching. One branch will see a live cat, the other will see a dead cat. *Both* of these will certainly happen. So there is no room for uncertainty here about what my future holds. I should expect to observe both a live cat and a dead cat.

Time and again, however, we do not *perceive* anything like branching. We perceive a single cat, alive or dead, and never both. So our experience is extremely hard to reconcile with this understanding of an Everett world.

So much the worse for the Everett view, claim the critics.

There is a second possible response, however. Rather than rejecting (P2), it is possible to disagree with the account of what is involved in imagining one or more possibilities. That is, we might have grounds to reject (P1).

Imagining egocentric possibilities

When we imagine two distinct possibilities, sometimes it is indeed clear that we are imagining two distinct ways for the world to be. For instance, we might imagine one way, in which Obama loses the 2008 election and John McCain becomes President of the United States. That possibility clearly corresponds to a distinct world from the one which we actually inhabit, where Obama was elected President in 2008.

But on other occasions, it is less clear that we are imagining distinct ways for the world to be. Imagine now that *you* are Barack Obama. Don't imagine that anything different happens to Barack Obama. And don't imagine that anything different happens to the body you now inhabit. Just suppose that you inhabit Obama's body. So you were born in Honolulu and raised largely by your grandmother. You studied as an undergraduate at Columbia and then went to Harvard to study law. You worked in Chicago as a community organiser before becoming first a state and then a federal senator. Then in 2008 you won the US presidential election. And so on.

What is the *content* of this imagined scenario? The assumption made above by those who criticised (P2) is that you are imagining a world that is physically just like our own, but that some additional ingredient – your 'soul' or 'identity' – has been relocated into Barack Obama. But this seems an unnecessarily cumbersome way of explicating what it is that you are imagining. Rather, you simply seem to be imagining *the same world*, but from a different perspective. Whether or not some other factor – an alleged soul – is located here or there is irrelevant.[6] What is relevant is just that *you occupy the body of Barack Obama*.

This suggests that (P1) is not true for all imaginings. *Some* imaginings do not involve imagining a different world, but merely involve imagining a different perspective on the same world.

The idea, then, is that the content of some imaginings involves not just an imagined world, but also a perspective on that world. We could represent this as a pair: $\langle W, E \rangle$. W is the world, or set of worlds, that captures the non-perspectival content of the imagining.[7] This content could be captured entirely in impersonal terms, such as:

There exists a person named Barack Obama and he becomes US President in the 2008 elections.

The other element in the content – 'E' for 'ego' – picks out a *location* as the perspective from which this content is imagined. Suppose W_a is the set of worlds that captures the content of *your* impersonal beliefs about the state of the world: your beliefs that Barack Obama is US President in 2010; that somebody, at some time, reads a book called *A Philosophical Guide to Chance*; that Shakespeare witnessed the first performance of *Twelfth Night* at one of the Inns of Court in London; etc. This does not capture all of your beliefs, however. You do not believe that *you* were present when *Twelfth Night* was first performed. You do not believe that *you* are the US President in 2010. And you *do* believe – I suspect – that you are currently reading a book called *A Philosophical Guide to Chance*. To capture this self-locating aspect of your beliefs, we need to specify *your location*. What is more, we need to specify your past and future location as well as your present one, because you believe yourself to be a thing with a past, present, and future. Your location is a path in space-time. It begins when and where you were

6 Caspar Hare (2009: 63–5) makes this point: arguing that identifying our judgments about personal identity with judgments about the persistence of a 'Cartesian Ego' fails to explain the array of possibilities we judge to be possible.

7 The idea that the content of a thought or belief could be represented by a set of worlds was introduced in §16.

born, passes through your current location, and ends when and where you die.[8]

Having noted that some of our beliefs and thoughts involve this self-locating content, it is evident that in some sense *much* of our thought is of this type. Any time we have thoughts about *here, there, now, the past,* or *the future,* we require a location from which those thoughts get their perspective. Think now about the thought you are having if you perform the observation in Schrödinger's famous thought experiment, and observe that the cat has survived. You might express the thought to yourself in just those words:

The cat has survived.

According to the Everettian understanding of what has happened, this way of speaking appears to be mistaken. There are now two cats, so referring to 'the cat', without further qualification, falsely implies that there is only one cat but does not uniquely pick out only one. But we often allow that, even without a sufficiently descriptive linguistic specification to identify a unique object, people may use the definite article correctly because the context makes clear what is being referred to. If I ask you to pass 'the salt', it is usually quite clear that there is a particular container of salt that I have in mind. Similarly, as an occupant of a branch with limited access to cats on other branches, my use of 'the cat' would be best understood as referring to the cat on my branch, and no other.

So provided we understand the thought that the cat has survived as a *self-locating* thought, then it will be possible to explain that thought as true, and it will also be possible to explain what it was that the observer was uncertain about. The observer on the live-cat branch should be understood as thinking:

I am located on a branch with a live cat.

And this observer, prior to the experiment, was unsure as to whether *this* would be true, or whether she would be located on a branch with a dead cat.

---

8 There will need to be some indeterminacy in the location you believe yourself to occupy, because you are unsure about some of the facts about where you have been and where you will be: Where exactly were you during your flight from San Francisco to Siberia? Did you pass over Alaska or not? At what time, precisely, did you come into existence? Where will you be at this time next year? Where and when will you die? This indeterminacy could be captured by taking a *set* of determinate paths through space-time. All paths in the set converge at those places where you are certain of your location, and diverge at various points where you are unsure. This account was developed in seminal work by John Perry (1979) and David Lewis (1979a).

So by understanding what seem to be non-egocentric thoughts about the outcomes of experiments as self-locating thoughts about being co-located with particular branch outcomes, we can recover distinct thoughts to associate with each experimental outcome. Although in some sense it is true to say that 'all possible outcomes of an experiment happen', it is also true, in a different sense, to say that 'only one experimental outcome will occur'. This second thought is true in the self-locating sense: *I will be located with only one experimental outcome.*

Uncertainty prior to the experiment

The proposal sketched above faces a difficulty, however. Here is the complaint, voiced by David Lewis.

There may well be uncertainty, and hence a subjective probability distribution, after a branching. A branch may not know which of all your branches it is – just as, when you are 'beamed up' to two different starships, each of your two arriving selves may at first be uncertain whether it is aboard the *Enterprise* or the *Potemkin*. This is uncertainty not about how the world is, but about who and when and where in the world one is. It is a probability distribution over alternative egocentric possibilities, not over possible worlds. But this genuine uncertainty comes after your branching, whereas your divided expectations for the future come before, when there was not yet anything for you to be uncertain about.   (Lewis 2004: 14–15 n.)

Lewis is complaining that the method I have sketched above, to identify two different possibilities that we may be uncertain about, only works after the experiment has been run. But we are unsure *before* the experiment has been performed whether we will see a live cat or a dead cat. And at this stage we have not undergone branching. There is only one person having this thought. So Lewis thinks we have equal claim to both branches. If we knew ourselves to be living in an Everettian universe, we should be certain that we will be located on *both* branches.

In making this objection, Lewis is apparently driven by some idiosyncratic assumptions about the nature of self-locating thoughts. Rather than proposing – as I understand it – that a self-locating thought is entertained by an entire person, with a past, present, and future, Lewis assumes that a self-locating thought should be understood as the thought of a 'person-stage'. A person-stage is simply a limited portion of a person, fixed in a particular moment of time. Before the experiment, at t_0, there is only one person-stage thinking about the possible outcomes. So when this person-stage entertains the two possibilities:

1. I will be located on the dead-cat branch
2. I will be located on the live-cat branch,

there is nothing about the person-stage that makes one of these true while the other is false. If we allowed there to be some factor which makes one true while the other is false, then we would be committed to claiming that there is some further fact about personal identity that would make it true that the person-stage survives down one branch or the other. On the physical facts alone, there is no basis for distinguishing which future branch belongs to the earlier stage.

So Lewis concludes that the person-stage must be correct to have *both* of these thoughts. Another possibility is that the person-stage is *incorrect* to have either of these thoughts. But whichever way we understand it, insofar as the *stage* is thinking about these things, the stage must have the same self-locating attitude to both future branches. Consequently, the person-stage has nothing to be uncertain about prior to the experiment.

But Lewis is incorrect, I suggest, to interpret these self-locating thoughts as the thoughts of person-stages. Rather, they are better understood as thoughts of *persons*. And there are two persons involved in the branching process, even before the branching has happened. There is the person who ends up on the live-cat branch and the person who ends up on the dead-cat branch.[9] Prior to branching, these two persons are co-located in the same body. See Figure 10.1 for an illustration of the situation. To keep separate the various candidate thinkers, we will call the continuant person who ends up on *Potemkin* 'Lefty', the continuant person who ends up on *Enterprise* 'Righty', and the person-stage that initially steps into the teletransporter 'Pre-transport DP'.

To motivate the thought that there really are two persons present, prior to the branching, imagine Righty, standing on *Enterprise*. He thinks to himself: 'I was wrong to think I would arrive on *Potemkin*.' The other teletransported version, Lefty, standing on *Potemkin*, thinks 'I was correct to think I would arrive here on *Potemkin*.' If Lewis is correct, both Lefty and Righty should interpret the earlier utterance of DP as a claim by 'Pre-transport DP' – their earlier, *shared* person-stage; and in that case, both Lefty and Righty should have the same attitude to this single thought: Pre-transport DP's thought was either correct or incorrect. But if we allow that the thought is the thought of a continuant person, then we can say there were *two separate*

9 Curiously, Lewis himself advocates this idea of co-location to best explain what happens in fission cases (1976b). But he fails to see how it could be helpfully applied to Everettian quantum mechanics.

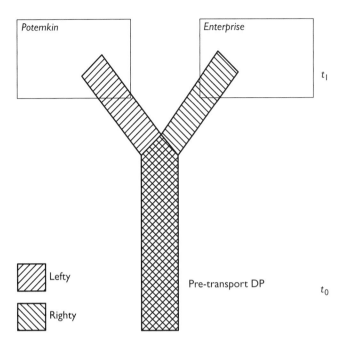

Figure 10.1 The two ships, understood as involving co-location. The early stage, 'Pre-transport DP' is occupied by two continuant persons, 'Lefty' and 'Righty'. Lefty goes to *Potemkin*, and Righty goes to *Enterprise*. (Illustration adapted from Lewis (1983b), with kind permission of Oxford University Press.)

thoughts being entertained at that time: one correct, one incorrect. Lefty and Righty were using the same body to entertain different thoughts. Lefty was thinking 'Lefty will go to *Potemkin*', and Righty was thinking 'Righty will go to *Potemkin*.' And, as we know, only one of those thoughts was correct.

The fact that there were two thoughts is concealed by the fact that the two thoughts are expressed using the same words or concepts, at the same time, in the same body. Pre-transport DP utters 'I am going to *Potemkin*.' But no one is to know – at this time – that Pre-transport DP is in fact occupied by two persons. The fact that there were two persons, hence two thoughts, does not become apparent until after the malfunctioning teletransportation brings about fission.[10]

10 This proposal was developed independently by Ninan (2009) in the context of personal identity and the semantics of first-person thought, and by Simon Saunders and David Wallace (2008) who were explicitly concerned to develop the proposal to assist in understanding Everettian quantum mechanics.

What does all this achieve? Provided we interpret *all* our ordinary talk about the future – claims such as 'the cat will live' – as involving self-locating claims about which branches we anticipate being located on, we can recover plenty to be uncertain about in an Everettian universe. Although I know for sure that *someone* will see the cat live, I do not know whether that will be *me*.

A very appealing feature of this account is that it emphatically vindicates the compatibility of determinism and chance. Our uncertainty about the future would remain, *even if* we had perfect knowledge of the present, the laws of nature were strictly deterministic, and we had unlimited computational power to deduce the implications of the laws. Our uncertainty is ineliminable because the propositions we are interested in are self-locating propositions. And self-locating propositions are irreducible to third-personal content.[11] The laws of nature simply determine what will happen in third-personal terms, so the laws cannot determine the propositions that we are interested in. Moreover, by claiming that self-locating propositions are irreducible to third-personal propositions, we are not claiming that physics is incomplete or that there is some mysterious non-natural phenomenon to be explained. Rather, we are simply exploiting the fact that there are some forms of meaningful thoughts and utterances that get their meaning, in part, from the perspective of the thinker. So this attempt to explain the role for uncertainty in an Everett universe elegantly deploys a minimal semantic tool to explain what might otherwise have seemed deeply mysterious.

Death in a branching world

Clever and elegant though it is, the above proposal runs into some serious difficulties.

Suppose I am sure that I live in an Everett universe, and I am trying to decide whether to purchase an insurance policy against house fires. The policy I am offered has two components: pre-death and post-death. If I purchase only the pre-death policy, then the house is only insured for fires that occur prior to my death. The post-death policy insures those who inherit my house for fires that occur after I have died.

Because quantum branching is ubiquitous, I am quite sure that there are branches in the future where the house will burn down, and there are also branches in the future where the house will be safe. My decision whether to

---

11 This has been persuasively argued for by Perry (1977; 1979) as well as Lewis (1979a).

purchase insurance is a decision about which of these branches I fancy that I will be located on.

Suppose I decide that the pre-death policy is a good deal and purchase it. I then consider the post-death policy. I find myself confused. After I am dead, there will be branches on which the house burns and there will be branches on which it does not. Earlier, I dealt with the multiplicity of branches by trying to work out on which branch I would be located. But after I am dead, I will not be located anywhere! So this would seem to suggest a very rapid argument for an implausibly strong conclusion: that I should not buy the post-death insurance at any price, because I will not be co-located with any house fire after my death. This result, if valid, would generalise to suggest that I should not care about anything that happens after my death.

Perhaps, less radically, I should conclude that I will be located on all of the post-death branches, since I am located on the 'trunk' from which all of those branches spring. But in that case, we seem to have struck again the very problem that was circumvented by the device of self-locating thoughts. There is no longer anything to be uncertain about: I should be certain that all possibilities will eventuate. The house will burn down and the house will not burn down.

Posthumous events

The posthumous events objection came to my attention through work by Huw Price (2010) and Alastair Wilson (2011). Wilson discusses two possible ways to try to repair EQM. The first proposal is that we can individuate persons, not by appeal simply to the continuant person, but by appeal to a maximal continuant – an entire world history, in effect. So when I am wondering whether to purchase the post-death insurance, I am wondering, of the many persons who will share my entire life history, which of them *I* am – am I one of the persons who is associated with a complete branch in which my house burns, and hence purchasing the insurance is a good deal, or am I one of the persons associated with a complete branch in which my house does not burn? There is no fact about us *during our lives* which accounts for these distinct identities. It is simply a brute and unanalysable fact that there are some persons associated with burned house branches and others associated with intact house branches.

This proposal runs into some internal difficulties in explaining what it is that things are made of, since all things turn out to be made of both ordinary material parts and also out of associated world branches (see Wilson 2011: 372–3). But even setting those aside, this proposal – as an attempt to explain what it is that we are uncertain about when we wonder what the future holds – does strain credulity somewhat. There was at least something natural about the thought that we might be concerned about where we will be located, where our location is a straightforward matter of physical locatedness. But this proposal makes our location a matter of brute association between our physical self and branch history that lasts long after our physical self has died. It is hard to see why this is something we have good reason to care about.

The second – and Wilson's favoured – proposal is to eschew the entire account of an Everett universe as a branching universe. Instead, he proposes that we understand the Everettian world as a *pluriverse* of diverging worlds. What we conceived as branches in our original presentation should be thought of as worlds that are numerically distinct, but happen to be qualitatively identical in the early stages of their history. Having started out similarly, they later diverge.

While this seems to have more luck in recovering our intuitions about what we should care about – of course we should care about our world to the exclusion of all others, even others that are qualitatively identical – it suffers from interfering with our intuitions about causation and distinctness of worlds. As discussed in §40, quantum mechanics involves interference effects that are most naturally explicated in terms of interference between world branches. This is held to be one of the great advantages of the Everett interpretation by some of its defenders. If we now maintain that separate branches are in fact entirely distinct worlds, then we must either treat quantum interference phenomena as causally inexplicable or we must allow that there is causal interaction between distinct worlds. Neither of these is an attractive option, I suggest.

In a recent paper, Simon Saunders (2010), supporting Wilson's line of thought, has claimed that 'there is no good reason to think EQM is really a theory of overlapping worlds' (2010: 202). He bases this claim on some weak mereological assumptions about how to apply the part–whole notion to the ontology of EQM. Saunders goes on to say, however, that to describe this as divergence 'is infelicitous . . . "Divergence" (inappropriately in EQM) suggests the absence of physical relations.' Notwithstanding Saunders' mereological claims, I take the existence of 'physical relations' to be excellent grounds for thinking that branches or worlds are in some sense 'in contact', or overlapping.

Given that we do not think it rational to treat the decision to purchase post-death insurance as radically different from the decision to purchase pre-death insurance, this suggests that the account which has been developed to explain uncertainty in an Everett universe is inadequate.

The problematic interaction between death and the Everett interpretation does not end here. It also suggests a suspiciously easy method of getting rich. All I need to do is find a bookie willing to sell me a bet that pays off a very large sum only if I survive an extremely risky quantum branching event. How will the bet turn out? My uncertainty, on this account, is uncertainty as to whether:

- I will be located on the branch where I survive. (The bet wins.)
- I will be located on the branch where I die. (The bet loses.)

But of course I won't be located on the branch where I die! I *cannot* be located on the branch where I die. So although a branch will exist where one of my descendants dies and the bookie does well, *I* certainly won't be located there. So I can be absolutely certain of success. I should be willing to stake any amount to purchase this bet.

More generally, for any physical scenario involving death, there will exist a bizarre branch involving extremely unlikely, but possible, quantum events which result in my survival. So I should never expect to die (Lewis 2004). If I am confident of the Everett interpretation, and if what I care about is what happens on branches where I am located, then I should be happy to walk in front of a bus, confident that *I* will only be located on branches where I live. There is bound to be some branch where an extraordinarily unlikely conglomeration of quantum events occurs, such that the bus particles pass through my body without disrupting any of my own particles.

(In case the reader finds this bizarre implication a welcome one, however, note that the argument does not suggest we are destined for immortality in good health, but rather are doomed to a future in which we *cannot* die, despite gradually accumulating more and more unpleasant afflictions (Lewis 2004: 20–1).)

Something has gone very wrong here. At best, it might be possible to retreat as follows: there is something very attractive about the thought that, if we live in an Everett universe, our ordinary thoughts should be interpreted as self-locating. But there is something literally incredible about supposing that we only *care* about self-locating propositions. So although the semantic strategies discussed above may shed some light on how to *think* about an Everett world, they should not be trusted to explain what we *care* about.

So Stage A is a partial success, at best. While it shows that we can recover a familiar notion of uncertainty for the events that occur in our lifetime, it does not readily translate into a way of thinking about events that risk death. And since all events bring some risk of death, this is a rather disabling problem.

But perhaps we can forgive this partial failure of Stage A. After all, death is a notoriously difficult event to understand. Ever since Epicurus argued that death is nothing to us, philosophers have been puzzling over how best to explain the misfortune of dying. Little wonder it remains puzzling when we think about it in the context of a bizarre world of parallel branches. What is really important is that we succeed with Stage B.

Stage B, recall, is the attempt to explain how particular degrees of belief should be rationally ascribed to particular branch outcomes. In particular we need to explain why the probabilistic Born rule works so well to predict and explain the apparently chancy events we observe. Think again of the story of the two ships. While we might now agree that it is legitimate to be uncertain about whether we will find ourselves, after the teletransportation, on *Potemkin* or *Enterprise*, it is hard to see on what basis we could assign a particular probability to one or the other outcome.

44 Indifference and branches

One thought which naturally occurs to people, upon introduction to the idea of a branching world, is that we should divide our beliefs evenly among the branches. If there is one live-cat branch, and one dead-cat branch, then surely the probability that we will see the cat live is simply ½?

This suggestion fails comprehensively.

As I noted above (§42), the idea of world branches is a metaphor that has some serious limitations. In particular, there is no intelligible way to *count* branches in an Everett universe, except perhaps to say that there are infinitely many branches of every variety. So simple ratios are mathematically impotent to prescribe what to believe (compare the difficulties raised for similar approaches in §22). What is needed is a *measure* over branches.

The Born rule provides us with that required measure: but it is not the only measure that we might use. So how can we justify our ascribing of probabilities in that particular fashion, using that particular rule? What makes that approach correct?

In response to this challenge, an argument has been produced by David Deutsch, a physicist based at Oxford University. That argument has subsequently been explored and defended in much greater detail by David Wallace, also at Oxford. The argument purports to show that we are rationally required to prescribe our beliefs in accordance with the Born rule.[12]

Suppose you are offered the opportunity to play a game, in which you have the choice between two boxes. One box contains nothing. The other box contains $100. Whichever box you choose, you can keep the contents.

In thinking about opportunities to play games like this, involving uncertainty, those in the subjectivist tradition of probability theory have shown how, from information about your preferences to play different games, we can deduce information about what you believe the probabilities to be of success and failure within the games. For instance, if you prefer to play the pick-a-box game versus playing a game in which you are certain to receive $80, then it looks as though you rate your probability of picking the box that contains the $100 very highly: much more highly than 0.5. The information I have given here is not enough to conclusively establish this conclusion, but it suffices to illustrate the basic idea: preferences for different sorts of games, involving different sorts of choice under uncertainty, determine information about probabilistic beliefs.

Suppose that it is rationally *required* that you have certain sorts of relationships between your preferences for certain sorts of games. That is, failure to have preferences of this particular sort would be evidence that you are irrational. Call the required relationship among a rational agent's preferences R, and call the preferences of an agent A, Pref_A.

For all rational agents, A, Pref_A possesses the property R.

Now suppose that, from the mere fact that an agent's preferences possess R, we can deduce some information about the probability A ascribes to a particular outcome of a choice situation. Suppose, for instance, we could deduce that an agent whose preferences possess R will think the probability of picking the box containing $100 in the above game is *equal* to the probability of picking the box containing nothing.

If this sort of deduction is possible, then it is seemingly possible to derive information about chance. For chance, recall, is approximately what it is best to believe, given the available evidence. If there are requirements for our credences that follow, merely from having rational preferences over games involving uncertainty, then surely these requirements are part of what it is best to believe – regardless of what evidence is available. It would be very

---

12 Deutsch's argument is presented in Deutsch (1999). Wallace has written a number of papers on the argument, but his most comprehensive treatment is in Wallace (forthcoming).

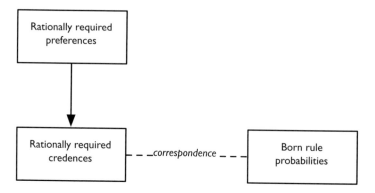

Figure 10.2 The decision-theoretic programme to explain probabilities in Everettian quantum mechanics.

surprising if rationality required us to believe something less than the best, given the available evidence.

For instance, if it were indeed possible to derive, in this fashion, that it is rationally required to believe that the probability of picking the box containing $100 is equal to the probability of picking the box containing nothing, then it is relatively straightforward to show that it is irrational to ascribe a probability greater than 0.5 to picking the box containing the money. If that were rationally permissible, it would lead to violating the probability calculus. Reasoning of this sort can get us to numerically precise credences, provided the initial relationship R among preferences is sufficiently demanding.

So this is the idea underlying Deutsch and Wallace's strategy to justify the Born rule:

1. Posit that certain preference relationships are rationally required in the face of a choice between different quantum states with uncertain outcomes.
2. Derive from these required preference relationships the corresponding requirements on our credences.
3. Show that the requirements on our credences include conformity with the Born rule.

See Figure 10.2 for a schematic illustration of the programme.

This strategy is certainly one that *can* work, provided one puts in sufficiently strong posits at the first step. But it is notoriously difficult to show that there are any rational requirements on our preferences. So much of the controversy over the Deutsch–Wallace programme relates to this point.

Table 10.1 The two games compared on what seem to be all relevant bases

Game A	Game B
One branch gets prize	One branch gets prize
Prize branch has spin up	Prize branch has spin down
Prize branch has amplitude $\frac{1}{\sqrt{2}}$	Prize branch has amplitude $\frac{1}{\sqrt{2}}$

Rationally required indifference

Suppose you were told that you could play one of two games, Game A or Game B. The description of each game is as follows:

Game A: A quantum branching event will occur: an electron will have spin up on one branch, and spin down on the other. You will undergo fission. Your fission product on the spin-up branch will receive a prize and the other will not receive a prize. The branches have equal amplitude.

Game B: This game is just the same as Game A, except that the prize will now go to your fission product who is on the spin-down branch, and no prize will go to the fission product on the spin-up branch. Again, the branches have equal amplitude. See Table 10.1 for a summary.

Now, let's suppose that you do not care whether or not you see spin up or spin down. What matters to you is *receiving the prize*, not how you get the prize.

There does not seem to be much basis, then, for preferring one rather than the other of these games. The only difference is a difference that we have stipulated that you do not care about.

Helping themselves to some assumptions from decision theory, Deutsch and Wallace argue that if we concede that it is rationally required for an agent such as this to be indifferent between these games, then an agent is required to adopt credences which conform to the Born rule! So, for instance, having shown that an agent must be indifferent between Games A and B, then we can *prove* that an agent must *not* be indifferent between Games A′ and B′, where these are just the same as Games A and B, but with unequal amplitudes.[13]

---

13 The literature on this topic is rather technical and I do not pretend to have given a complete treatment. For readers wishing to pursue this idea further, Hilary Greaves (2007) has written a valuable survey article, which is probably the best place to begin. Readers wishing to delve deeper are advised to examine Wallace (forthcoming).

This might seem rather suspicious. How can showing that we must be indifferent between A and B lead to a conclusion about required discrimination between other games?

For instance, consider another variation on Game A, which we'll call Game A-minus. In this game, there is *no prize on either branch*. One of the decision-theoretic axioms ('Dominance') *requires* us to believe that it is *less probable* that we'll get a prize in A-minus than in A. Given that there are no prizes in A-minus, that sounds pretty uncontroversial! But recall that this claim about probabilities is supposed to derive from a rational requirement on our preferences. We are *rationally required to prefer* Game A over Game A-minus, given Dominance, and given that we prefer a prize over no prize. That is less plausible. In the fission picture, we should think of Game A-minus as akin to creating an identical twin of ourselves, and doing nothing more. In Game A, similarly, a twin is created, but one of myself and my twin gets a prize, and the other does not. With the complication of identical-twin creation added in, it is not clear that I am rationally required to have such preferences. I might, for instance, be strongly committed to achieving egalitarian outcomes for myself and my twin, and hence prefer Game A-minus. Or I might be extremely fatalistic about such games, assuming I will always be on the losing branch, no matter what. In that case, I might remain indifferent between Game A and Game A-minus.[14]

It is crucial, then, for the success of the Deutsch–Wallace approach, that it be appropriate to apply these decision-theoretic axioms to a novel sort of decision: a decision over multiple branches rather than among different possible outcomes, only one of which will be actual.

Even supposing that Deutsch and Wallace can reassure us of the rationality of their axioms in this novel situation, there is a residual reason why we might remain suspicious of the Deutsch–Wallace strategy. In effect, their strategy shows us that a physical symmetry requires of us a certain corresponding sort of symmetry in the distribution of our credences over branches. It remains deeply mysterious why a physical symmetry alone should be enough to make this rationally warranted. The most successful explanations of how we infer chances from symmetries make appeal to our knowledge of dynamical laws and other empirical discoveries about the nature of the world.[15] The Deutsch–Wallace approach does not seem to do anything analogous. So it

---

14 See Price (2010) for a more thorough presentation of this concern. And see Greaves and
    Myrvold (2010: 296–301) for some attempt to meet it.
15 See Strevens 1998 and also the discussion above, in §29.

Table 10.2 The two games compared on what seem to be all relevant bases

Game C	Game D
One branch gets prize	One branch gets prize
Prize branch occupant is fat	Prize branch occupant is thin
Prize branch has amplitude $\frac{1}{\sqrt{2}}$	Prize branch has amplitude $\frac{1}{\sqrt{2}}$

remains suspiciously 'magical' that it can conjure a chance out of a pure physical symmetry.

Alternative measures?

One further criticism that has been made of the Deutsch–Wallace programme is to try to show that, when faced with a choice between quantum branching events, it might be rationally permissible to accept a rule other than the Born rule. David Albert (2010), for instance, has suggested that it seems perfectly rationally acceptable to apportion my credences, not merely in accordance with the amplitude squared measure, but also in accordance with the degree to which my branch descendants are fat, the whimsical rationale being: there is 'more of me' on branches where I am fat, so I should care more about what happens there.

So consider the choice between Games C and D (see Table 10.2). In both of these games, due to quantum branching events, I will have a descendant on one branch who is fat and a descendant on the other branch who is thin. In Game C, the fat-branch descendant gets the prize. In Game D, the thin-branch descendant gets the prize.

An agent who adopts Albert's fatness measure will seemingly be in disagreement with Deutsch and Wallace about the probability of winning a prize. Albert will have preferences which correspond to being more confident that he will win the prize in Game C, where he is fat on the prize branch, than in Game D, where he is thin on the prize branch. Deutsch and Wallace want to argue that we should be governed by the amplitude alone in setting our degree of confidence. But it is not clear why it is *irrational* to have preferences of the sort Albert describes.

That said, Albert's pattern of preferences is decidedly odd, as can be seen by thinking through an example in more detail. Suppose that Albert continually contrives to experience quantum branching events in which he will be thinner on one branch and fatter on the other. The amplitude squared measure of these branches is always equal, and given what we know about the observed outcomes of experiments like this, Albert will therefore 'find himself' on the thin branch just as often as he finds himself on the fat branch. Quite often, then, Albert will have to console himself that although his present thin self missed out on a prize, at least his fat self has one.

Usually, we would take experiences like this to be evidence for or against a probabilistic hypothesis. If we keep betting a certain way, and 'losing', we revise the betting strategy. It is hard to see how Albert could confirm or disconfirm any of his beliefs about the fatness measure. Should he take himself to have learned that relative fatness does not impact upon probability in the way he was seemingly committed to at the beginning? Or could Albert rightly refuse to change his probabilistic beliefs, because after all, they are simply grounded in his preferences. Albert could look back on all the many trials in the past and say: 'I have "found myself" on the less important branch in every trial. But that is not to say that my judgments of importance were incorrect. They were correct. The outcomes on those other branches *are* more important, even though I am not located on them!'

The point of this issue is that we need to do more than think about preferences over games in a single case. We also need to think about how we revise our preferences in a way to permit us to learn from experience. That is exactly what has been done by some more recent defenders of an Everettian approach.

45 Bayesian learning about branches

The most promising recent development of the decision-theoretic programme to explain probabilities in an Everett world has been advocated by Hilary Greaves and Wayne Myrvold (2010). Greaves and Myrvold have tried to show that they can explain why rational agents would adopt the Born rule, making fewer controversial assumptions than Deutsch and Wallace.

First, Greaves and Myrvold take a terminological stand to try to minimise confusion about how it is that probability can be applied to a deterministic theory such as Everettian quantum mechanics. They say they are not going to provide a theory of chance, but rather will explain how a different quantity, *branch weight*, plays a similar role in our epistemology as chance. And they will show how good epistemic practice in our beliefs about branch weights will lead us – given the sorts of evidence we have seen – to adopt the Born rule as the best account of the weight of various branches.

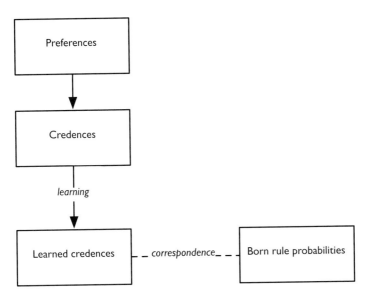

Figure 10.3 The Greaves–Myrvold account of how we might vindicate the Born rule in an Everett universe.

Having made that terminological distinction, Greaves and Myrvold's main innovation over Deutsch and Wallace is to show how, without presupposing that we are rationally required to adopt any particular preferences between different branching outcomes, a Bayesian story about how we update our credences can be told – but in this case, it is a story about branch weights, rather than about chances. And given the great deal of evidence which shows that quantum experiments give rise to distinctive frequencies of outcomes, the application of this Bayesian updating will lead us to adopt the Born rule as a well-confirmed hypothesis about branch weights. See Figure 10.3 for a simple illustration of the Greaves–Myrvold approach, as opposed to the Deutsch–Wallace approach.

Suppose we start with someone who does not treat Game A and Game B as equally desirable, and suppose that – contra Deutsch and Wallace – this is not irrational. Then the credences of these people will fail to conform to the Born rule. But that is not a problem, say Greaves and Myrvold, because agents who revise their credences in the light of experience will nonetheless, over time, be rationally compelled to adopt Born rule credences – or credences extremely close to them. (At least, this is going to be the case for agents on branches like ours, with frequencies like ours. There will also be some 'unlucky' agents who are exposed to frequencies that disconfirm the Born rule.)

The Deutsch–Wallace argument seems like magic. From a bit of information about the games, it conjures up a precise probabilistic rule that seems to be more than we could ever have deduced on quasi a priori grounds. Greaves and Myrvold, by putting more explanatory burden on an account of how we update our beliefs about chances from experience, avoid this suspicion. But note: Greaves and Myrvold also assume that decision-theoretic axioms such as Dominance can be applied to a branching world. So their approach ultimately inherits many of the same objections that can be levelled at the Deutsch–Wallace approach.

But before dwelling on the problems, let's consider in more detail how the account works.

Recall the account given, in §37, of the psychological aspects of an agent who believes in chance and takes experience to confirm or disconfirm chance hypotheses. This account was built up from the following elements:

- A link between preferences and credences, such that a sufficiently rich preference state determines a credal state.
- An account of *conditional* preferences, which can be interpreted as conditional probabilities. Moreover, if an agent's conditional preferences are effective, then the agent will be engaged in updating her credences by conditionalisation.
- An account of the distinctive attitude which an agent must have to the data, in order to be treating the data as relevant to an objective chance. The crucial point is that the agent must regard *frequency* information as important – she must care about how many times, out of the total, the coin lands heads – but must also be insensitive to the *order* in which the data is collected: she doesn't change her view depending on whether all the heads came at the beginning, at the end, or were interspersed throughout.

Note that there has been no mention of 'uncertainty' in this account. The only way uncertainty might enter is by being a presumed feature of the *wagers* over which an agent has preferences. The whole story can be given, however, without the assumption that the wagers involve any uncertainty. Instead, the choice between wagers could be a choice between branching events, where we are *absolutely sure* that both outcomes will happen. Provided we have a sufficiently rich set of preferences over these branching events, and that we meet the other constraints loosely described above, then we will have mental states that are just like credences in chance hypotheses, and we will have a process of updating our preferences over branching events that is just like updating credences in those chance hypotheses by conditionalisation.

So return to our hypothetical agent from Chapter 8 who had preferences over the following games:

Ice-cream game   The player simply gets some ice-cream.

Cake game   The player simply gets some cake.

Heads-is-ice-cream   A coin is tossed. If heads, the player gets ice-cream. If tails, the player gets cake.

Heads-is-cake   A coin is tossed. If heads, the player gets cake. If tails, the player gets ice-cream.

Red-ball-is-ice-cream   The player draws a ball from an opaque barrel containing red and black balls. If a red ball is drawn, the player gets ice-cream. If black is drawn, the player gets cake.

This player has the following preferences over these games:

- Ice-cream game > Cake game.
- Heads-is-ice-cream $\simeq$ Heads-is-cake (the player is indifferent between these games).
- Red-ball-is-ice-cream > Heads-is-ice-cream.

But now, understand these games as involving *quantum branching*. So the game Heads-is-cake is understood as a game in which the world will divide into one branch where there is a heads event and one branch where there is a tails event. The player's heads descendant will receive cake; her tails descendant will receive ice-cream. Similarly, in all the other games. Instead of the player being uncertain what will occur, she is certain that both possible outcomes will actually happen. Greaves and Myrvold stress that, even if she *knows* that this is what will happen, and even if she is completely certain about what the future holds, *she could still have the same preference structures as she had in the non-branching case.*

Moreover, claim Greaves and Myrvold, the rationality constraints that decision theory imposes on preference structures in normal cases involving uncertainty seem similarly plausible in the branching case. If that is right, then agents have a preference state that seems to suffice for a mental state that is analogous to *credence*. This is the sort of assimilation of preference to credence that Deutsch and Wallace make in the set-up to their proof.

But now that we have focused so heavily on the fact that these credence-like states do not necessarily involve any uncertainty, it seems strange to think of them as degrees of belief. After all, they are supposed to be mental states that we can possess, even with complete confidence in what will happen in future. Greaves and Myrvold suggest that the state which we ordinarily call a credence or degree of belief is functioning more like a degree of *care*. By being indifferent between the games Heads-is-cake and Heads-is-ice-cream, I am expressing my *equal degree of concern or care* for what happens to my heads-branch descendant and my tails-branch descendant.

By having non-dogmatic attitudes towards our branch descendants – by having degrees of care that depend on the 'outcomes' of branching events – we will be able to adjust our subjective degrees of care in a way analogous to revising our probabilistic beliefs. We will be engaged in much the same process as Bayesian updating of credences by conditionalisation, but here it will be Bayesian updating of *concern*.[16]

Provided we have the right sort of indifference to the sequence in which the branching outcomes are revealed to us, then we will behave like agents who are tracking an objective 'chance-like' quantity. In this case, we will follow Greaves and Myrvold in calling it branch weight, and informally characterise it as an objective measure of how much we *should care* about a given branch.

So Greaves and Myrvold have shown that an agent who has the right sort of preference structure – again a preference structure that looks prima facie rational, even in a branching world – can be represented as one who is entertaining various hypotheses about the branch weight of the various branches that she knows will come to exist in future. Her branch descendants, by adopting the preferences that her causal ancestor preferred, on discovering the outcome of experiments, will come to adopt branch-weight hypotheses that make the actual sequence of events more weighty (more likely). This process is structurally analogous to updating credences in chance hypotheses, and is similarly linked to dispositions to accept wagers, so there is some reason to think it is psychologically much the same too.

The upshot of all this is that Greaves and Myrvold have outlined some precise conditions under which epistemic agents in an Everett world would come to adopt the Born rule, if they had been exposed to evidence much like

---

16 Note that Greaves and Myrvold must appeal to the same sort of self-locating information that Saunders and Wallace employ so heavily, if they wish to make sense of the idea that we discover an 'outcome' of a branching event. We *know* that one descendant will experience heads and that another will experience tails. But we do not know where we will be located. It is this discovery about our own location that is supposed to trigger the revision of our degree of care.

For Saunders and Wallace the dependence on self-locating information was problematic, because it meant that probabilities for posthumous events did not appear to make much sense. This is less threatening for Greaves and Myrvold, however, since they only require the self-locating information for *learning*. Thus they can restrict their attention to propositions about where an agent is located at *present*, rather than the more confusing cases about future location. Moreover, since no learning occurs after death, there is no need to worry about making sense of our attitude towards posthumous events.

we had. They would do this because, having been exposed to frequencies that fit the Born rule very well – as we have been – they will come to adopt preference structures which are equivalent to having a high degree of credence in the Born rule chances.

46 Evaluating the Greaves–Myrvold account

What does this explain?

It is a slippery thing trying to get a grip on what this sort of account of scientific practice is actually explaining. Here is a rough attempt to get at what Greaves and Myrvold might think they have explained:

If the Everett interpretation were true and we knew it to be true, it is still possible that we would engage in something like normal scientific practice, and we would still have beliefs about chances that are much like our current beliefs. In order to have this happen, we would merely need to have certain preference structures over branching events. Moreover, these preference structures look quite close to structures we rationally aspire to in the actual world.

So if we *are* in an Everett world, although we are greatly mistaken in our belief about the indeterminacy of the future, we are not entirely mistaken to have credences in chance hypotheses. We need to reinterpret those credences somewhat, but there is no call to dispense with them.

Question: does this help to explain, in the way that alternative accounts of chance explain, why chance has the two crucial features that it does:

i. chance explains observed frequencies, and
ii. chance constrains rational credence?

It seems to give some explanation of the second: chance constrains rational credence, simply because we cannot help but be entertaining branch-weight hypotheses provided our preferences have a certain, rationally required, structure.

But it surely cannot explain the first. We *know*, if we know the Everett interpretation is correct, that every possible frequency of outcomes is actual. So how could a frequency of outcomes be explained? Do we seek an explanation of the *particular* frequency of events that we have seen? How can the weight of our branch explain that, unless we think that there is some mysterious extra-physical force by which we are made more likely to inhabit a high-weight branch? Do we seek an explanation of *all* the frequencies that are actual? We don't need branch weight for that: all the frequencies are already explained by the mechanism of the Schrödinger evolution, without any mention of how much we care about them.

The answer must be, I take it, that we do explain our particular branch frequencies by appeal to branch weights. And the thought is that agents *treat frequencies as evidence for branch weight*; therefore agents are committed to branch weight being explanatory of frequencies. That may be so, but we might think it is not what we wanted. We did not ask for an account of why agents living in an Everett world might *believe* that branch weights explain the frequencies they observe, but rather an account of what makes that belief *correct*.

Which assumption to reject?

Having illustrated in broad brushstrokes my concern that this proposal cannot explain frequencies in the way required, how would I diagnose where Greaves and Myrvold go wrong? Given they purport to offer a rigorous proof from certain assumptions, is one of their assumptions problematic?

I have already pointed out, in discussing the Deutsch–Wallace approach, the reasons to be suspicious about at least one vital decision-theoretic axiom: Dominance. If we take the reality of all branches seriously, then this axiom seems at odds with our normal way of thinking about how to care for future individuals. We normally think of future individuals as separate beings who must be treated as equals, at least in some important regards. The axiom of Dominance leads us to take an attitude that differentiates between individuals in a way that is potentially disturbing, or at the very least not rationally required.

But I want to focus on the assumptions that Greaves and Myrvold use to generate the account of updating by conditionalisation. There are two axioms they propose: 'P7' and 'P8', in their nomenclature. P7 is the requirement that an agent will revise her preferences in accordance with her conditional preferences formed before an experiment.

P7. During pure learning experiences, the agent adopts the strategy of updating preferences between wagers that, on her current preferences, she ranks highest.

Unless we actually *have* alternative strategies for updating preferences, P7 won't have any bite. Thus Greaves and Myrvold postulate an assumption of non-dogmatism (P8). Using traditional terminology of chance and credence, it is the idea that an agent should not adopt zero credence with respect to any chance hypotheses that could explain the data.

That sounds fair enough, given that adopting a credence of zero with respect to a hypothesis is much like treating it as a logical contradiction. But loosely translated into the language of care, it says this:

P8. An agent should not have dogmatic preferences about wagers: she should allow that her preferences might change – and might change to practically any configuration – on the basis of experience.

There is nothing pre-eminently rational about this!

P7 and P8 are reasonable-looking requirements in the original framework of chance. If we are talking about chances, an agent has a prior belief that one outcome out of many will happen, but it is not certain which possible outcome. So upon experiencing something that is actual, one might be getting evidence of what is more likely. If one adopts dogmatic credences, then one forecloses on the possibility of learning from these experiments. An agent like that won't be able to do science. This seems bad.

But in the branching case, an agent who has initial degrees of care about outcomes has little or no reason to hedge his or her bets across different chance hypotheses. Nothing is learned by the revelation that the electron has spin up on *this branch*, except some self-locating information. Why should I change my preferences in light of that? Why should I have adopted hedged preference strategies, such that I have different future preferences in light of different self-location discoveries?

The only reason why I would do such a thing, it seems, is if I think that facts about my location reveal something about the importance of a branch. 'I tend to be located where the serious stuff is happening.' I call this sort of thought: *care depends on where*.

That we must be committed to this thought, as a *requirement of rationality*, is highly dubious.

Someone could reasonably reject this attitude, by pointing out that 'all branches are going to happen anyway, so why should I think I have learned anything about the weight of a branch, just because I find myself on that branch? So whatever my initial branch-weight hypotheses, they can never be changed by experience.'

More idiosyncratically, it also seems possible that I could accept P8, but reject P7, in the following way. I could accept the idea that my location reveals something about the importance of branches, but I might be convinced that I tend to be located on *less* important branches. This would lead to some sort of counterinductive strategy, analogous to someone who infers from observing a high frequency of p, that p is very *un*likely. Stated as a claim about chance, this seems absurd. But as an attitude to adopt for updating preferences it does not seem to be ruled out by a requirement of rationality.

What remains of chance?

Ingenious though the above ideas are, none are sufficient to show that there are objective chances, corresponding to the Born rule probabilities, in an Everett universe. These accounts fail for two reasons. First, they struggle to show that we are *rationally required* to adopt any particular credal state, given the available evidence. So they cannot vindicate the idea that a chance is the uniquely best advice that can be given, given available evidence. Second, they fail to show how chances can explain the frequencies we actually see, for reasons that are familiar from traditional criticisms of subjectivism.

But despite being a failure, the failure of the Everett programme is an instructive one. It shows us that, if the world is as described by the Everettians, it is not surprising that we have turned out to have a concept of chance. Bear in mind that, if the Everett view is correct, there will be a staggering plenitude of branches. In many ways, these branches will resemble an array of possible worlds: all sorts of things we took to be merely possible will be real in some branch or other. Unlike other celebrated theories of possible worlds, however, the physical laws will be the *same* in all branches (whereas David Lewis, for instance, argued that there are possible worlds where the laws are otherwise). But given the nature of the laws of quantum physics, this will still allow branches where very strange things happen. Indeed, we should expect branches where very different galaxies exist, branches where life evolves in many more places, and branches which do not evolve life-forms at all.

Given the diversity of branches, there will exist some branches in which creatures will possess the sorts of psychology described by the Everettians. We can hazard a rough description of the sorts of conditions under which this psychology will tend to be fitness-enhancing, and hence selected for in evolutionary processes. Of course, there will also be other branches where other cognitive strategies will do better.

For instance, creatures who live in extremely simple branches, where the statistical patterns in the salient events appear to be governed by extremely simple rules, might be able to get by with innate beliefs, rather than the cognitive flexibility required for Bayesian conditionalisation. And creatures who live in perverse, counterinductive worlds might find that the benefits of Bayesian conditionalisation are too small to be worth the cost. Other organisms with more haphazard cognitive facilities might out-compete them.

These are two sorts of exception. There surely are others. But even so, it seems plausible that there will remain a very wide range of branches in which creatures evolve that approximate the psychological constraints required for Bayesian epistemology. In all cases, these creatures will seemingly be committed to *care depends on where.* Of course, by sharing this commitment, but being located in different places, they have different views about the correct branch-weight hypotheses. For instance, some organisms, exposed to frequencies that are close to those predicted by the Born rule, will advocate the Born rule as the correct branch-weight hypothesis. Some, in branches with different frequencies, will advocate different hypotheses.

What should we think about this disagreement? In one sense it appears to be a case of *no-fault* disagreement, like that between two individuals who have normative commitments that are completely alien to one another, and hence unable to agree on any normative propositions. In that case, it is hard to see how chance talk could be understood in a realist fashion, for if there were a fact about chance, then these disagreements should be capable of being settled one way or another.

On the other hand, it is tempting to think that *both* types of organisms are guilty of error in an important way. They both think that facts about branch weight explain the frequencies that they have seen in the past. But if the Everett hypothesis is correct, branch weight does not explain in any objective sense. A complete explanation of the frequencies that are actual is given by the Schrödinger equation. Beyond that, the only explanation of why we see the *particular frequencies* on the *particular branch* we inhabit is a deflating one: 'it had to happen to somebody'. This failure of branch weights to explain, then, seems to favour an error-theoretic interpretation of chance talk.

So, having taken a close look at the ways chance might be explained in quantum mechanics, we have come up with much the same result as was reached in Chapter 8: the most attractive accounts are anti-realist. I will return to the case for anti-realism about chance in Chapter 12. For now, I merely want to conclude that, as feared, the Everett hypothesis is incompatible with a realist conception of chance. Whether that is itself a reason to reject the Everett hypothesis is a further question that I leave untouched.

11 | Time and evidence

47 The time asymmetry of chance

If there are real chances, then it seems as though there are 'more' of them, in some sense, in the future. To illustrate, take some events which are governed by stochastic laws. For instance, at the tip of my nose, there is an atom of carbon-14, the radioactive form of carbon. Being radioactive, there is some chance that the atom will undergo radioactive decay. The typical type of decay for carbon-14 is known as *beta-minus decay*. This is a process whereby an atom emits two particles: an electron and an electron anti-neutrino, and in addition one of the atom's neutrons is converted to a proton. The result, when this occurs in an atom of carbon-14, is that it becomes an atom of nitrogen-14.

Whether this particular atom of carbon-14 will decay in the next year is a matter of chance. The chance is very small, but greater than zero. In contrast, for the year that has just *passed*, it seems that there is nothing chancy about what happened to the atom. The atom of carbon-14 'got here' in one particular way – though we may not know what that is – and there is no chance that it in fact developed in any different way. So there is something chancy about the future of this atom, but the corresponding fact about its past does not seem to be chancy. This may lead us to ask: where does this asymmetry come from? Why is the past different from the future, regarding the distribution of chances?

We might think – at least for chances involving the decay of radioactive isotopes – that the asymmetry arises from the law governing decay itself. The law tells us that the *future* is chancy, but it appears to be *silent* about the past. So, perhaps we have an asymmetry in the distribution of chances in virtue of an asymmetry in the law.

This way of speaking about the laws of radioactive decay is certainly very common, but if pressed, physicists will reveal that it is not strictly correct. First, we need to note that in addition to the process of beta-minus decay of carbon-14 to nitrogen-14, there is a physically possible process that we might call 'spontaneous constitution' of carbon-14 from nitrogen-14. This process is just the *time-reverse* of the beta-minus decay process,

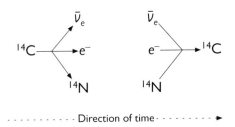

Figure 11.1 The process of carbon-14 decay and its time-reversed process.

in the very same sense that was discussed in §10. In the time-reverse of beta-minus decay, three particles *come together*: an atom of nitrogen-14, an electron, and an electron anti-neutrino. And these three particles interact in such a way that the electron and anti-neutrino are absorbed, and one of the nitrogen protons is converted to a neutron. As a result, the three particles are converted into an atom of carbon-14. See Figure 11.1 for an illustration.

The *dynamical* laws of nuclear physics – the laws that tell us how protons and neutrons and the like interact – simply tell us something like this: given you have an atom of carbon-14 *now*, if you run the clock *forwards* by a given amount, then there is a particular chance that, by the end of that time, the atom will have undergone decay. But – and here is the surprising part – the law also says that, if you run the clock *backwards* by a given amount, then there is the very same particular chance that since the beginning of that time, the atom will have been created by spontaneous constitution. In effect, the law is indifferent to the direction of time in which the process occurs. The law simply tells us that, given we have an atom of carbon-14 now, the chance that a decay/constitution event occurs in an adjacent period of time equals a certain value.[1]

So as far as the dynamical laws are concerned, if we know nothing more than that there is an atom of carbon-14 now, there *is* a reason to think that there is a chancy fact about its past. That is, the law tells us, given there is a carbon-14 atom now, there is a chance that, in the *last year*, an atom of nitrogen-14 underwent a process of spontaneously absorbing an electron and an electron anti-neutrino, and thereby formed the atom of carbon-14 we now see. That chance will have the very same value as the chance of decay in the future.

---

1 Note that the chance that a given carbon-14 atom underwent spontaneous constitution in the past hour is *not* the same thing as the chance that an atom of nitrogen-14 will participate in spontaneous constitution in the next hour.

We now have a real puzzle. What the law is telling us is symmetrical with respect to time, and it is directly at odds with what we think to be true about chance. It seems just plain *wrong* to say that the carbon-14 atom has a chance of having spontaneously formed from nitrogen-14 in the last year.[2] Instead of believing that there is a chance regarding the atom's past, it seems more plausible to believe:

- the chance is zero. It simply did not happen, so it has no chance of having happened. Or,
- the chance is one. Amazingly it *did* happen.

Or, given that we are not in a position to tell which of these is true, we might wish to distribute our confidence between these two possibilities:

- Either the chance is one or it is zero, but we do not know which.

Finally, if we are more open to chanciness for past events, as I have recommended in earlier chapters, then we might agree that there is a non-trivial chance that the atom was spontaneously constituted, but we know that spontaneous constitution of carbon-14 from nitrogen-14 simply *does not happen*. Or if it happens, it is with *staggeringly* less frequency than the time-reversed process – the process of beta-minus decay. So this difference in the frequencies must be reflected in the chances. So we might claim that:

- The chance that the atom was spontaneously constituted in the past year is non-trivial, but is much lower than the non-trivial chance that the atom will decay in the next year.

Even if you find some of these views implausible, I take it that at least some of them are much more plausible than the claim that the chance of past constitution is *equal* to the chance of future decay. So *some* view on which the past and future chances are asymmetric is surely most plausible.

This is the same puzzle we touched on briefly in §21. The laws of physics are frequently time-symmetric. This means that the laws entail that certain processes that occur in one direction in time could occur in the opposite temporal direction. But for many processes, we never see anything like this occurring, and our ordinary experience gives us very good reason to think that time-reversed processes *cannot* occur. Eggs cannot be unbroken.

---

2 As it happens, carbon-14 *is* typically formed from nitrogen-14, but *not* by the time-reverse of beta-minus decay. Rather, the typical process of carbon-14 formation involves nitrogen-14 being bombarded with neutrons.

Aging cannot be reversed. Light cannot be absorbed back into a torch. And similarly, it seems that carbon-14 cannot be spontaneously constituted.

So why do the chances appear to display a time asymmetry that is not present directly in the laws? Given the way I suggested we characterise chance – as the best thing to believe, given the available evidence – the time asymmetry of chance is most obviously explained by an asymmetry in the *availability of evidence.*

Consider the sort of evidence we have of the past. We have fossils, which give us good evidence for the evolutionary history of the Earth. We have memories, which may be fallible but are still enormously useful sources of information about our own past. We have photographs, which give us good evidence for how various historical figures looked. We have diaries, newspapers, audio- and video-recordings. We have polar ice-cores, whose carbon dioxide content is recorded to give us a better understanding of the climatic and atmospheric conditions of the planet hundreds or even thousands of years ago. And we have the traces of light and other radiation that reach us from space. Cosmologists have been able to use these to form reasonable beliefs about the very earliest moments of the universe, billions of years before the present.

This sort of evidence – the evidence of traces, records, and memories – is obviously very important to our ability to find out so much about the past. It is difficult to imagine just how ignorant we would be if we lost all of that information.

Now think about the sort of evidence we have for the future. We have no memories of the future. Occasionally, individuals claim some sort of foresight, but there is little reason to think these claims true. Even if they were, there would still be much *less* evidence available by means of foresight than by means of memory. We also have no records of the future. There are no diaries or photographs brought back to us by time travellers. And there seem to be no traces of the future. There are no physical phenomena that can be used to infer how the world will be in the future, in the way that layers of ice from the polar regions contain traces of the carbon dioxide levels which inform us about conditions in the past. There are no footprints from the future. There are no patterns of light arriving from stars in the future. And so on.

Corresponding to this paucity of records, traces, and memories of the future, we also find that we are in a position to *know* much less about the future. We pay various experts to make predictions about – for example – the weather, the economy, politics, and the ecosystem. But these predictions are extremely imprecise and unreliable compared to what we know about the past.

Above I mentioned that it is hard to imagine just how ignorant we would be if we lost all of our records, memories, and traces of the past. Having realised, though, just how little we know about the future, we get a good idea of what the effect would be for our knowledge of the past if we lost all those sources of information. We would have to make *retrodictions*, in much the way that weather forecasters and the like make *predictions*. We would take our best understanding of the current state of the world and, using some model, we would extrapolate to some earlier state, hoping that this extrapolation will be a reasonably accurate estimate of the past.

In light of this, it seems obviously true that regarding typical future facts there is *much less evidence available* than for typical past facts. Returning to the puzzle about the atom of carbon-14, we now have a genuinely temporal asymmetry to point to, so as to explain the asymmetry of chance. Recall my characterisation of chance as what is best to believe, given the available evidence. Let's assume the laws are part of the available evidence. The chance for the atom's past is different from the chance for its future, not because the laws dictate that the future for a carbon-14 atom is different from its past, but rather because there is a relatively great amount of evidence available regarding the past, it is frequently the case that the best thing to believe regarding a past-proposition is to give it credence one or zero. For propositions about the future, this is not the case. There are frequently propositions about the future for which the best available advice is to keep an open mind. *It is because there is a lot more available evidence regarding the past than there is regarding the future that the past is less chancy than in the future.* That, in short, is my proposal to explain the asymmetry of chances.

For something as small and difficult to notice as an atom, this appeal to an asymmetry of evidence might seem puzzling. We have not been making records of the carbon-14 atom's state in the past. We have no memories of its earlier state. So is there really an abundance of available evidence about its past?

The evidence may not be abundant, but there is certainly *more* evidence about its past than about its future, for two reasons. First, regarding the atom itself, there could well be different traces left in the world, depending upon whether it was an atom of carbon-14 for all of the past year or whether it was spontaneously created from an atom of nitrogen-14. If it was carbon-14, it may well have behaved differently in the presence of light or other radiation. The light it emitted or absorbed could have struck a photograph, or could still be travelling through space. In some fashion, then, the earlier state of the atom could have 'left a mark'. And if a trace of the atom's former state exists, then there is some evidence *available* regarding its earlier state, even if no one ever perceives that trace.[3]

Second, even if – remarkably – no mark has been left by this particular atom's earlier state, we have a lot of good inductive evidence that atoms of nitrogen-14 only very rarely undergo spontaneous constitution of carbon-14. So there are good inductive grounds for thinking that no such thing has happened to this atom. Regarding the future of the atom, however, experience does show that it certainly can and does happen that atoms of carbon-14 decay into nitrogen-14. So we have much less reason to be sure that decay won't occur in the future.

So we have a proposal to explain the temporal asymmetry of chance. Even where the laws are symmetrical in their chance implications about past and future, we still may have asymmetric chances, due to an imbalance in the available evidence. Much more evidence is available about the past than is available for the future.

Some readers may already have some objections to this account. I will try to address those later. First, our explanation invites a further question: why is there an asymmetry in the available evidence? Why, if the world is governed by laws that are – by and large – indifferent to the direction of time, do we see some processes occurring only in one direction, and not the other? Why do we see traces left behind of the past, but never receive portents of the future?

48 Explaining the asymmetry of evidence

I have suggested that it is the abundance of traces, memories, and records of the past, and the paucity of anything similar for the future, that contributes to the uneven distribution of chances with respect to time. The future is more chancy than the past, we might say. Having made that suggestion, it would certainly be desirable to say something about why there is a relative

---

3 My claim that past microscopic events leave traces more than future microscopic events is to some extent controversial. David Lewis attempted to use an asymmetry of this sort to explain the temporal asymmetry of counterfactuals (Lewis 1979b), and the view has been criticised as implausible for counterfactuals regarding microscopic events by Price (1996: 148–50). See also Frisch (2010: 21) on the tendency for traces to 'wash out'. I address this concern in the main text, in §50.

abundance of traces, memories, and records of the past, and a relative paucity of such rich evidence regarding the future.

The causal hypothesis

One obvious feature in common across traces, memories, and records is that for a trace T to be direct evidence of some phenomenon P, T must be a *causal effect* of the earlier phenomenon P. The process by which we get traces, memories, and records is always a causal one, and in the cases where traces are direct evidence, it is always a process that goes *from* the phenomenon as cause *to* the trace as effect.[4]

If we were to have traces, records, or memories of *future* events, then we would seemingly require there to be causal processes that run from *later* phenomena to *earlier* traces. That is, we would require there to be causation which runs 'backwards' in time. And backwards causation appears to be non-existent or extremely rare. We ordinarily never observe or posit causal relations from the future to the past or present, and trying to imagine backwards causation usually involves imagining very strange scenarios, such as those described in time-travel stories.

There may be some truth in this explanation. But it is not entirely satisfying. Why is there an asymmetry of causation? Why is it that it seems either impossible, or at least extraordinarily rare, for future events to cause past ones? Again, given the symmetry of the physical laws, it is hard to see why this should be the case.

The counterfactual hypothesis

One very plausible thought is that there is an asymmetry of causation because of an asymmetry of *dependence*. We seem to understand causation through its intimate connection with our reasoning about what depends on what. Claims about dependence are often expressed using counterfactual conditionals (discussed in §18). If I claim that the pollution in the river caused the failure of the crops, and you asked me to explain what I meant

---

4 There are also cases where we obtain evidence that P has occurred because we observe a causal effect of some phenomenon that is typically also a cause of P. So, for instance, if we hear thunder, this is indirect evidence that it is raining, not because rain causes thunder, but because both thunder and rain are frequently caused together by certain sorts of atmospheric conditions. Addressing this sort of case explicitly will lead to needless complication without disrupting the basic idea that causal connections are necessary for the existence of good evidence, so I will ignore it henceforth.

by that, I might say: 'Had the river not been polluted, the crops would not have failed.'

This claim about the way the crop failure depends upon the pollution is probably not going to succeed as an *analysis* of causal talk, but it is clearly picking out something very similar to the original causal claim.[5] Counterfactual claims, like the claim that the failure was dependent on the pollution, are often produced as *evidence* for causal claims.

Counterfactual dependence claims are – typically – temporally asymmetric. We think that later events depend upon earlier events, but not vice versa. So perhaps the temporal asymmetry of causation is explained by this temporal asymmetry of counterfactual dependence.

But of course, because counterfactual dependence is so intimately tied to causation, to explain the temporal asymmetry of causation by appeal to a temporal asymmetry in counterfactual dependence is not really to advance us much further. Why should either of these relations – the relation of counterfactual dependence or the relation of causation – be temporally asymmetric?

The causal laws, again

One idea – which we examined earlier – is that the temporal asymmetry of causation might arise as a result of an asymmetry in the causal laws. Perhaps the laws by which the world evolves have a direct temporal asymmetry built into them, such that the lawful evolution of the world in some sense 'proceeds' from earlier states to later states, and not vice versa. For instance, laws might have the logical form:

If a system is in state P_1 at an earlier time, that system must be in state P_2 one second later.

If the laws had this sort of structure, then there would still be *some* lawful constraints upon earlier times, given later times. For instance, we could draw inferences such as: 'The system is not in P_2 now, so the laws entail that the world was not in P_1 one second ago.' But if the states P_1 and P_2 are complete, then this is obviously a much weaker constraint than the constraint this law imposes in the forwards temporal direction. Being told that the world *is not*

---

5 Much effort has been spent on trying to analyse causation in terms of counterfactuals, largely inspired by David Lewis's seminal paper, 'Causation' (1973a). For a useful anthology of more recent papers that largely revolve around this research programme, see Collins, Hall, and Paul 2004.

in a given, fully determinate state is much less informative than being told that it *is* in such a state.

But we have already seen that proposals like this run into immediate difficulty: they seem to be contradicted by what we know of physics. As already noted in §10, the laws are – for the most part – invariant under time-reversal. Some laws specify functional relationships between variables that hold at a given time and other laws specify ways that systems evolve over time. But in neither of these sorts of laws do we see any reference to one direction of time rather than another. Apart from a few minor exceptions, there is no interesting temporal asymmetry in the laws.[6]

A more subtle suggestion can be made, however, in defence of temporal asymmetry in the laws. Tim Maudlin, for instance, does not deny that the equations we use to describe physical laws are invariant under time-reversal. But Maudlin claims that even though the equations show no preference for one or the other temporal direction, it is very plausible to think that the *world itself* undergoes change and evolution in one temporal direction only, and that this fact is intimately related to the nature of the laws. The idea is that, even if a mathematical formulation of a law does not explicitly mention temporal direction, it is a further fact about the world that there is an objective *passage* of time from earlier to later states – and that this further fact is why we can say that present states *produce* later states, in accordance with the laws, but not vice versa. This is the source of a deep physical asymmetry.

What reason does Maudlin have to posit this additional fact about the passage of time? He writes:

If *all* that physics ever appealed to in providing accounts of physical phenomena were the dynamical laws, then we would seem to have a straightforward Ockham's razor argument here. But the laws of nature *alone* suffice to explain almost nothing. After all, it is consistent with the laws of physics that the universe be a vacuum, or pass from the Big Bang to the Big Crunch in less time than it takes to form the stars. The models of fundamental physical laws are infinitely varied, and the only facts that those laws *alone* could account for are facts shared in common by all the models. In all practical cases, we explain things physically not merely by invoking the laws, but also by invoking *boundary conditions*. So to argue that the direction of time plays no ineliminable role in physics demands not only a demonstration that no such direction is needed to state the laws, but that no such direction plays any role in our treatment of boundary conditions either. (Maudlin 2007a: 120)

Maudlin goes on to argue that the way we use boundary conditions in certain types of physical explanations displays a temporal bias. It is legitimate to cite an early state of a system to explain a later state. But it is not similarly legitimate to cite a later state of a system to explain an earlier state. For instance, if I am asked to explain why the room was so messy at the end of the day, it would be perfectly acceptable to cite the fact that there were a number of small children visiting during the middle of the day. It would be very odd to explain the presence of small children in the middle of the day by citing the need for something to cause the mess at the end of the day!

Certainly, we do discriminate between possible explanations in this way. But it is very difficult to establish that the use of earlier times to explain later times reflects a metaphysical fact about the passage of time. What would it be like to live in a universe in which much the same particular events occur as in this world, but in which the passage of time runs the other way? Would we be able to tell, if we lived in such a world, that explanation should run in the opposite direction? (See Loewer (forthcoming) and Price (2011) for further discussion of difficulties relating to Maudlin's account.)

An alternative account of our explanatory practices is possible. Perhaps it is because of an asymmetry in the sorts of things that we know that we try to explain later states in terms of earlier states. That is: perhaps the asymmetry of evidence itself can be used to explain the asymmetry of explanation. I am optimistic that this is a better strategy than Maudlin's but I will leave the argument for another occasion.

The thermodynamic hypothesis

We have an asymmetry of available evidence, an asymmetry of causation, an asymmetry of chances, and an asymmetry of counterfactual dependence. None of these asymmetries appears to be directly explained by the fundamental, dynamical laws of physics. But there is a branch of physics that includes one profoundly important asymmetric law: thermodynamics.

Thermodynamics is the branch of physics that deals with how *heat* – a particular form of potential energy – changes its form over time. The second law of thermodynamics, which is the famously time-asymmetric law, says that the heat that is available for work in a closed system always remains the same or decreases in the future. Or, to put it in its most familiar formulation: *entropy never decreases.*

This slogan, however, is not very enlightening without a better grasp of the meaning of terms like 'entropy' and 'heat'. The science of thermodynamics emerged largely from the attempt to build better steam engines, and consequently the concepts most naturally apply to systems of gases at various pressures and temperatures in chambers with pistons and the like. In that context, for the uninitiated in the physics, it is tempting to think of heat as a noun that corresponds to the adjective 'hot'. The degree of heat is just the degree of 'hotness', or temperature. But this temptation must be resisted. The concept of heat is importantly different from the concept of temperature. Rather, heat means something like *the potential energy that is absorbed or lost by a system when it increases or decreases in temperature.* So exchange of heat occurs, for example, when you put some ice in a liquid, the ice melts and the liquid cools.

Although heat is primarily understood as a form of energy that manifests in changes of temperature, it can also be converted into *mechanical work.* Imagine a chamber of gas that contains a piston. The gas is under high temperature and pressure, and because of this, the piston is pushed out. The movement of the piston is mechanical work, and because the gas has given up heat to do this work, the temperature of the gas goes down. (This is a good illustration of the first law of thermodynamics, which is a conservation law. It states that the total amount of energy – be it in the form of heat or kinetic energy – remains constant.)

Entropy is an even more difficult concept to explain in familiar terms. The entropy of a system is properly defined as the amount of heat in the system that is *un*available to do mechanical work. As it happens, there is a strong match between our intuitive notion of disorder and the definition of entropy. Roughly, when at the molecular level a system is highly ordered, then there is likely to be more heat available to do work. And, conversely, when at the molecular level there is more disorder, then relatively more of the system's thermal energy is unavailable for work. Think first of a balloon that has been inflated with helium in a room at normal pressure. Then compare this to a state where the balloon has been deflated and the helium molecules have dispersed evenly throughout the room. The second arrangement is intuitively much more disordered. The helium molecules are no longer neatly arranged in the balloon, but scattered everywhere. Corresponding to the difference in molecular order between these two states of the system, there is a difference in the entropy. When the balloon is inflated, the compressed helium contains heat, which can be used to propel the balloon about if we untie it. Or the helium can be used to make a loud noise if we pop the balloon. Once the helium has been released from the balloon, there is less heat available for useful work. So the change from a less disordered to a more disordered state corresponds with a change from low entropy to high entropy.

The second law, somewhat more carefully stated, appears as claims like:

- A cyclic transformation whose only final result is to transfer heat from a body at a given temperature to a body at a higher temperature is impossible.
- A cyclic transformation whose only final result is to transform heat extracted from a source which is at the same temperature throughout into work is impossible.
- The total entropy of any isolated system, in the course of any possible transformation, either keeps the same value, or goes up.[7]

Remarkably, these very different-looking formulations can – with a minimum of auxiliary assumptions – be *proven* to be equivalent (Albert 2000: 24–33). Having learned this, it is perhaps not surprising to be told that the second law has enormously broad ramifications. For instance, the second law in some sense explains why, if we put an apple in a closed box, it will decay rather than stay the same. Even though that seems to have nothing much to do with gases and steam engines, it can be shown to be a consequence of the second law.

If the second law of thermodynamics could be taken at face value, then we could reason as follows: when a number of atoms of carbon-14 in a closed system undergo beta-minus decay, entropy increases. The time-reverse of that process – spontaneous constitution – involves a decrease in entropy. So the second law states that carbon-14 *cannot* be spontaneously constituted in a closed system. This gives us a very powerful explanation of why there is an asymmetry of chances: there is *zero* chance of spontaneous constitution of carbon-14 (in a closed system), because if spontaneous constitution were to occur it would violate the second law. And, more generally, every instance we find where a process has a non-trivial chance in the future, but is the sort of thing we would never expect to have happened, time-reversed in the past, then there is good reason to think that the time-reversed process will be ruled as impossible by the second law.

---

7 The first of these is Clausius's formulation, the second is Kelvin's, and the third is the entropy formulation. Taken from the entry on 'Thermodynamics', *Encyclopaedia Britannica Online*: www.britannica.com/EBchecked/topic/591572/thermodynamics.

Unfortunately, things are a bit more complicated than this. As already discussed in Chapter 5, the second law of thermodynamics is now understood by physicists *not* as the sort of law that holds without exception. In fact, it is somewhat misleading to call it a 'law', as though it is on the same footing as the laws of quantum mechanics. Our best way of understanding the second law is as something like a very strong probabilistic generalisation. Entropy-lowering processes *can* happen, but they are extremely unlikely.

In earlier sections (§§21–23), we have examined the statistical reasoning that supports this way of understanding thermodynamics. In particular, I introduced a key element in a statistical–mechanical explanation: the statistical 'postulate'. This tells us that the probability that we are on a given type of trajectory through phase space is proportional to the 'volume' occupied by trajectories of that type. The appropriate volume measure over phase space tells us there is a high probability – indeed, *overwhelmingly* high – that we are on a trajectory that will move to a higher-entropy state.

So far so good. This is enough to tell us that, in future, things will probably develop in accordance with the second law. However, a problem which we noted at the time but did nothing to address is that the reasoning used so far is entirely time-symmetric. This gives rise to a very counterintuitive implication, and it is worth spelling out the reasoning for it rather slowly. The simplified statistical–mechanical explanation for why we should expect entropy to increase in the *future* is something like this:

F1. The world is currently in macroscopic state M (a non-equilibrium state).
F2. In the region of phase space that represents worlds in state M, the volume of trajectories which have increasing entropy in the future is *massively larger* than the volume of trajectories that have constant or decreasing entropy in the future.
F3. The probability that we are living in a world of a given type is proportional to the volume of that type in phase space, relative to the volume of the current macrostate.

Therefore:

F4. The probability that we are living in a world where entropy will increase in future is massively larger than the probability that we are living in a world where entropy stays constant or goes down in the future.

That is the rationale for it being extremely probable that entropy will increase *in the future.* To this end, the rationale is successful. But, unfortunately, the rationale proves too much. To see this, now assume that we do not know anything about the entropic state of the past. It is possible to use precisely this rationale – suitably time-reversed – to draw a similar conclusion about the past. The argument would run as follows:

F1. The world is currently in macroscopic state M (a non-equilibrium state).

P2. In the region of phase space that represents worlds in state M, the volume of trajectories that have increasing entropy towards the past is *massively larger* than the volume of trajectories that have constant or decreasing entropy in the past.

F3. The probability that we are living in a world of a given type is proportional to the volume of that type in phase space, relative to the volume of the current macrostate.

Therefore:

P4. The probability that we are living in a world where entropy increases towards the past is massively larger than the probability that we are living in a world where entropy stays constant or decreases in the past.

So this reasoning – if sound – establishes that in the *past* the world was in a much *higher*-entropy state than it currently is. But this conclusion is absurd. If we know anything at all about the past, then one of the things we know is that the world – or at least the part of it visible to us – was in a lower-entropy state. Indeed, for all our experience of the universe, entropy has been going up. And practically all of the things we have learned from cosmology, archaeology, and palaeontology presuppose that the world was in a very low-entropy state in the distant past. If we re-did those sciences of the past, on the assumption that entropy has decreased since then, we would get all sorts of crazy conclusions.

To get a feel for what I mean by 'crazy' conclusions, focus on the example of archaeology. Think about how our current civilisation – focusing on our architecture in particular – will endure in future years, after humans have gone extinct or otherwise stopped caring for the buildings. Given our common-sense knowledge – which is perfectly consistent with thermodynamics – we expect that marks and scratches will gradually begin to be left on the buildings from passing animals and from the weather. The windows will break. The paint will peel and the plaster will crack. Eventually, the tallest parts will fall off. The foundations will rot. In general, the buildings will *fall into disrepair*. The state of disrepair is a higher-entropy state than the state of being well maintained. It is extremely plausible to suppose that all this will happen in the future, absent some special intervention to prevent it from happening.

All of these claims are compatible with the above account of statistical mechanical reasoning. But if we accept the time-*reversible* account given so far, then it is similarly plausible to suppose that the *time-reverse* of these processes *happened in the past*. So, looking at a building *now*, it would be reasonable to infer that, in the distant past, the building was in a state of terrible disrepair. Through seemingly random movements at the molecular level, the disrepair was 'undone'. Leaks were repaired, windows were unbroken, peeling wall-paint spontaneously stuck to the walls, and rotten foundations gradually dried out and 'unrotted'. Perhaps most remarkable of all, pieces of brickwork that lay on the ground, through a random confluence of molecular movements, jumped up from the ground and settled in place in the walls. In other words, a sequence of events that we would regard as utterly miraculous would, by the time-symmetry of statistical mechanical reasoning, be the best explanation of the current state of our civilisation. This is manifestly absurd.

The way to get more plausible conclusions out of statistical mechanics is in some sense crude, but effective. Roughly, we revise (F3) to apply only at the very beginning of the universe, and to include the assumption that *the early universe was in a remarkably low-entropy state*. This assumption, that the universe was in a remarkably low-entropy state, has become known as the 'past hypothesis'.[8]

F3′. The probability that we are living in a world of a given type is proportional to its volume in phase space, relative to the volume of worlds which satisfy both: (i) macro-condition *M* at present and (ii) the past hypothesis in the distant past.

If we now try to reason about the past using premises F1, P2, and F3′, we find that we can no longer derive such ridiculous conclusions. Instead, we need to modify P2 to better fit with our premises and reason as follows:

F1. The world is currently in macroscopic state *M* (a non-equilibrium state).

P2′. In the region of phase space that represents worlds in state *M that also satisfy the past hypothesis*, the volume of trajectories that have decreasing entropy towards the past is *massively larger* than the volume of trajectories that have constant or increasing entropy in the past.

---

8 My exposition of this closely follows David Albert's (2000: 81–5), though Albert goes on to make further proposals about the nature of the probabilities involved. I wish to remain neutral on that controversy here.

F3′. The probability that we are living in a world of a given type is proportional to its volume in phase space, relative to the volume of worlds which satisfy both: (i) macro-condition *M* at present and (ii) *the past hypothesis.*

Therefore:

P4′. The probability that we are living in a world where entropy decreases towards the past is massively larger than the probability that we are living in a world where entropy stays constant or increases in the past.

Consistently with P4′, we are able to invoke lower-entropy states in the past to explain the phenomena we see today. Consistently with P4′, we can explain the traces of past civilisations as having come about in the sorts of ways that we already thought of as plausible.[9]

It is thought that this restriction of the probabilistic reasoning involved in statistical mechanics will suffice to explain all the time-asymmetric phenomena associated with the second law. So, in some sense, these two elements of statistical mechanics – the statistical postulate and the past hypothesis – ground the second law of thermodynamics. Note that, given this grounding, we do not get the result that the second law is strictly true. Rather we get the result that it is *overwhelmingly likely* to be true. For any system we consider, it is *overwhelmingly likely* to conform to the second law.

This asymmetry has good 'physical credentials' in that physicists really do agree that it is sufficient to explain the second law. Moreover, the asymmetry of the second law is a profound and widespread asymmetry. Unlike the few particles that appear to display time-asymmetric behaviour on the microscale – which many doubt can be involved in explaining the macroscopic time asymmetries we see – the second law asymmetry is extremely pervasive. It looks like a good candidate to explain lots of other asymmetries.

But that work now needs to be done. Why is it, if we live in a world where the second law of thermodynamics is true, that we have temporal asymmetries of causation, of evidence, of counterfactual dependence, and of chance? For our purposes, the main game is to explain the asymmetry

---

9 Note: I am not suggesting that, until we hit upon statistical mechanical reasoning and the past hypothesis, we were not justified in believing that past civilisations existed. Rather, until statistical mechanical reasoning was explicated in this way, it was *mysterious* how our understanding of microphysical processes could be reconciled with what we already knew about the past. This is primarily a story of improving our understanding of physics, rather than a story of how we improved our understanding of archaeology, history, palaeontology, and the like.

of chance. And, consequently, that means explaining the asymmetry of *evidence*.

49 Statistical mechanics and the temporal asymmetry of evidence

Take a piece of evidence that we consider to be informative about the past. On the wall of a building we see some writing that appears to have been made with an aerosol can of paint. It says 'Tom was here'. The paint is a bit faded in parts, and there are bits of dirt and grime on top of some it.

Prima facie, this is evidence that some time ago someone called Tom was present at this spot, and that he used a can of black spray paint to record his presence. Of course, it is far from decisive evidence that a person called Tom was at this spot. Maybe Mary was at the spot, and she wanted to make it look as though Tom was here, so she wrote the graffito to create that impression. We can imagine many other alternative explanations. In all plausible explanations, however, we will cite the existence of a can of black spray paint, and an earlier state of the wall when it was relatively more clean than it is now. We will also appeal to a process in which somebody or other used the can of paint to create the writing we now see. I take it that this is because the current state of the wall is compelling evidence for these things. It is conceivable that we are wrong to draw these conclusions, but it would take some extraordinary evidence to convince us of that.

All of this is obvious. What is not obvious is how we can employ statistical mechanical reasoning to explain why evidence like this – a relatively informative record of the past – is relatively common, while portents of the future are so rare.

Albert (2000: chap. 6) has argued that our use of records is akin to using a *measuring device*; and we obtain information from measuring devices as follows: first, we assume ourselves to know something about the state in which the measuring device begins the measurement. We assume that the thermometer is well calibrated, that the Geiger counter is switched on, that the stopwatch counter reads zero, etc. Call this state of the measuring device its *ready condition*. Our first assumption is that the measuring device is in its ready condition. Second, we assume ourselves to be able to know the outcome of the measurement. We look at the measuring device, at the end of the experiment, and we reliably form beliefs about whether the thermometer reads 99°C or 100°C; whether the Geiger counter needle points to 17 rads or 17.2; whether the stopwatch reads 1′00″42.1 or 1′00″45.1

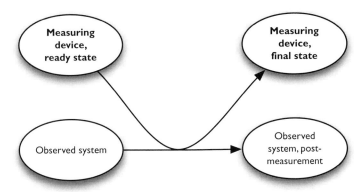

Figure 11.2 The process of measurement. From our knowledge of the measuring device's **ready state** and **final state**, we infer something about the state of the measured system at a time *in between* – at the time of measurement.

From this information that the measuring device was in its ready state, and that the measuring device is in a given final state, we can infer the state of the measured system at a time in between these times; see Figure 11.2.

Our knowledge of the ready state is essential for this inference to be reliable. If it is a serious possibility that the stopwatch was not reset before we began the experiment, then we cannot be confident that we have measured duration accurately. If the Geiger counter has a needle that sticks, and won't let it get past 17 rads, then we cannot be confident that the measured system is not much more radioactive. If the thermometer does not work properly, so it is permanently stuck at 99 °C, then we have no reason to conclude anything at all about the temperature of the measured system.

Similar assumptions underlie our use of the mark on the wall as evidence that a person was present at this spot, and that the person used a can of spray paint to mark the wall. We are assuming that, in the past, the wall was in a state in which there was *no black spray paint on the wall*: in other words, that the wall was *clean* in the past. If we were presented with compelling evidence that the wall was, in the past, covered in much *more* black paint, then we will have to posit some other explanation to account for the smaller amount of black paint now. We are also assuming that the molecules of black paint that are now spread out over the wall were once *collected together* in an aerosol can. If we were informed that this was never the case – indeed, that the molecules were never more collected together than they are now – then again, our original explanation becomes vastly improbable, and we will be obliged to adopt an alternative.

This sort of inference, from an earlier ready state and a final state to an intermediate state, is enormously powerful and widespread. It seems to be what is at work in all our uses of records. And note how little we can learn about the thing we measure *without* assuming anything about the ready state. Imagine you find a stopwatch which reads 32.4 seconds. Assume you know the macroscopic state of the watch, the dynamical laws, and the statistical postulate, but you do not know the past hypothesis. What could you conclude about the recent past from this? Albert suggests that the best inference you could make would be a sort of retrodiction. That is, you would be compelled to make the same sorts of inferences as we normally make about *the future*. What is most likely to happen to this stopwatch in the future? It is overwhelmingly likely – eventually – to deteriorate and break. It is possible that the stopwatch may interact with other systems in the future, such that it will measure them. But you have *no reason to think that its present reading is indicative of the state of those future systems*. Similarly, if you cannot assume anything about the past of the stopwatch, *you have no reason to think that its present reading is indicative of the state of those past systems*. So the stopwatch reading becomes useless as a record of the past.

Albert's observations about measuring devices seem very plausible. But we have not yet said anything about entropy or about the second law. What does this have to do with the entropic gradient of the universe, and what does it have to do with the ubiquity of records of the past and the relative scarcity of portents of the future?

First, we can note that the sort of information we have, when we know that the measuring device was in its ready condition, and when we know the final state of the measuring device, is information about the world's macro-condition. Second, what we are trying to determine, in ordinary measurements, is another fact about the world's macro-condition: the macrostate of the observed system at the time of measurement. So there is sure to be a statistical mechanical story to be told about this sort of inference. That is, using these two macrostates, and using our knowledge of the dynamical laws and the standard probability measure over phase space, we should in principle be able to say what is the probability that the world was in a given macro-condition at an intermediate time. Of course, this is too difficult for us to actually do. We do not know the exact macro-condition of the entire world: we merely know, approximately, the macrostate of a small region of the world. We do not have enough compu-tational power to work out the particular trajectories that pass through the regions of phase space that represent our knowledge: we merely generalise, based on our knowledge of the dynamics about what the typical trajectory will be like. So Albert is not assuming that we actually use this statistical mechanical account in our ordinary practice of measurement. He assumes, rather, that this is a plausible *maximal limit* on what we can know by measurement.

Using this idea of a maximum amount that we can learn by a given technique, Albert's point is that if we do not allow ourselves knowledge of the measuring device's ready condition – even if we make extremely generous assumptions about how much we know about other things – then it still will not be possible to draw the sorts of inferences about the past that we tend, in fact, to draw. So in order to draw these inferences about the past from records, we must be *presupposing* something about the earlier condition of the recording devices. If that presupposition is correct, our use of records will be reliable. If the presupposition is incorrect, our use of records will be extremely unreliable. Indeed, it will be about as reliable as using the current reading on a stopwatch to predict the duration of a future process.

How does this idea – that we are presupposing a ready condition in our use of recording devices – bear on the second law of thermodynamics?

For many of the traces and records that we use, we were not around to ascertain the ready condition of the device that took the recording. We listen to an old LP of a symphony orchestra, in which we hear an oboe make a mistake in the third bar of the second movement. This is surely good evidence that a very particular event happened: that an oboist in that particular orchestra made that particular error while performing that particular piece on that particular occasion. But we were not there to set up the phonographic recording. What is our evidence that the phonographic recorder was in its proper ready condition?

Suppose we tried to find some evidence. Suppose we interviewed the audio technician who was there, and he asseverates that it most certainly was in its ready condition, that the recorder was in good working order, and so forth. Now we are trying to use the technician's *memory* as a record of the recording equipment being in its ready condition. But to use the technician's memory like this, we need to assume that the technician himself was in his 'ready condition'. Was he an honest observer, unhindered by drugs or distraction? We have no direct evidence about that either. No matter what other evidence we hope to identify, given that it is *present* evidence of the *past*, we will again be in a position of having to assume something about the ready state of the relevant 'measuring device'. That will push us back to a yet further assumption about an earlier time, and it is clear that we have embarked on an alarming regress.

Reflecting on these sorts of cases brings us to realise that, in treating something as a record of the distant past, we are unavoidably *assuming* something about the condition of the universe prior to the record being made. There are two questions to be asked here: *what is it* that we are assuming? And *how do we know* that our assumption is correct?

Let's focus on the first question. Albert claims that, in all cases, *one* common assumption will do the work required. That assumption is simply that the earliest observable parts of the universe were in a very low-entropy condition. In other words, our assumption is that the past hypothesis is true. As Albert puts it, the past hypothesis is 'the mother . . . of all ready conditions' (2000: 118).[10]

It is not easy to follow Albert's argument that the past hypothesis *must* be the claim that serves this grounding role, but it is extremely suggestive, especially in the light of some examples. Consider the case of the black paint on the wall. The assumed earlier state in which the wall is clean and the paint is all in the spray can is a much lower-entropy state than the present condition. And, in general, it is very hard to see how we could construct a measuring device of any sort which ends up in a *lower*-entropy condition after a measurement than before.

What about the second question: how do we know that this assumption is correct? This is perhaps the wrong question to ask. Many epistemologists have thought that, if I know that P only if condition Q obtains, it does not follow that I must know that Q in order to know that P. All that is required is that Q does in fact obtain. Much the same might be true here. Using my memory of yesterday's weather as a record, I *know* that there was some precipitation that fell in my yard. This inference is only reliable if my brain was in its ready condition. I don't *know* that my brain was in its ready condition. But so what? I nonetheless know that there was no precipitation in the yard yesterday.

---

10 There is an intriguing analogy between Albert's past hypothesis proposal and Kantian ideas about the synthetic a priori. Kant argues, for instance, that it is a synthetic a priori truth that every event has a cause – because this is a precondition of our having any knowledge or thought at all. Albert's past hypothesis, similarly, is thought to be a necessary presupposition of our drawing powerful inferences about the past from records. Hence it is supported by the sort of transcendental reasoning favoured by Kant. But the past hypothesis, it should be stressed, is an a posteriori claim, and is therefore open to empirical defeat. So its epistemological status is much less furtive than any Kantian transcendental deduction.

Albert has argued that the past hypothesis grounds not only the asymmetry of evidence but also the asymmetry of control: the fact that we can seemingly affect the future but not the past (2000: 125–30). Drawing further on this idea, Barry Loewer (2007) has tried to explain the temporal asymmetry of counterfactuals in terms of the past hypothesis. These proposals have met with some serious criticisms and difficulties (Frisch 2005, 2007, 2010; Price and Weslake 2010). The rough idea seems to show great promise, but it is very difficult to tie something as remote as the past hypothesis into a precise account of something so strikingly *local* as our notions of control and counterfactual possibility. Another difficulty relates to the fact that the temporal asymmetry of control seems to be *strict*: we do not have *any* control over the past. It is not merely that we have much *more* control over the future than the past. But the Albert–Loewer proposals invite the consequence that we do have some control over the past – sometimes to a counterintuitively large degree (Frisch 2010).

The solution to the puzzle is now nearly complete. We are trying to explain why the universe contains ubiquitous records of the past, but seemingly contains no portents of the future. If Albert is correct, the existence of records of the past requires the past hypothesis to be true. So, conversely, if there were to be portents of the future, it would require that a *future* hypothesis be true.

But there is no good reason to think that a future hypothesis is true. There is no evidence that the world is going to end up in a low-entropy state. To see why this is so, reflect on what sort of circumstances would give us reason to accept a future hypothesis. For the future entropy to be very low, at some point it will have to begin *decreasing*. If this started to happen, we would see, for instance, ice cubes forming spontaneously in glasses of water, with the unfrozen water heating up in the process. And this is just the beginning: the world would be full of processes that we would have normally regarded as utterly anomalous. (But if we came to accept that this sort of thing was normal, then there would indeed be portents of the future in such a world. A glass of water which contains a half-melted ice cube in it would be a portent of a future in which the ice cube will be larger and the water warmer.)

So, to bring this all back to the question with which we began – why is there an asymmetry of chances? – I suggested it was because of an asymmetry of evidence. Why is there an asymmetry of evidence? Following Albert, I suggest it is due to the fact that, in our universe, a past hypothesis is true, while no analogous future hypothesis is true. Perhaps there is even further explanation possible: perhaps there is something to be said about why the world has a remarkably low-entropy past. But for our purposes this explanation will have to suffice.[11]

50 The roles of evidence, availability, and context

The idea that the low-entropy past is involved in explaining time-asymmetric phenomena is widely accepted. What is peculiar and relatively novel about my proposal is that I stress the evidential status of the low-entropy past. It is because the past hypothesis is something we *know*, or because the past hypothesis underwrites our *ability to know* other things, that the low-entropy past of the universe is part of our explanation of the asymmetry of chances.

My reference to the epistemic status of the past hypothesis might appear to the reader to be redundant. Isn't the past hypothesis *itself* enough to explain the asymmetry? Why is it necessary to mention our ability to know this or any related facts?

Consider a simplified example, which is reasonably analogous to the case of carbon-14 decay.[12] Suppose we have some robots that are designed so that, after every minute has passed, there is a 50 per cent probability that the robot will move one metre to the left, and a 50 per cent probability that it will move one metre to the right. The dynamical law governing the behaviour of these robots is effectively a chancy law, and the law is *symmetrical* regarding whether the robots move left or right.

You show me a robot, but you do not show me where in the room it is. You ask: what is the chance that it will move right in one minute's time? The answer is easy: 50 per cent. You ask: what is the chance that it will move left in one minute's time? Again: 50 per cent.

But suppose that I tell you that all the robots I have been showing you were initially placed to the far left of the room. If a robot is next to the leftmost wall and tries to move left, nothing will happen (and similarly if it tries to move right at the rightmost wall). I also tell you that the room is very large, and that not much time has passed since I first switched the robots on. So you infer that the room would look something like the illustration

11 See Price (2004) and Callender (2004) for opposing views on whether we should seek any
    further explanation of the past hypothesis.
12 Kindly suggested to me by Huw Price.

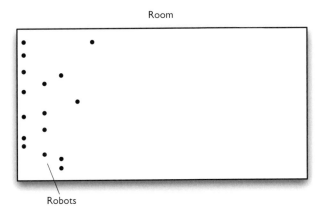

Room

Robots

Figure 11.3 This represents a likely distribution of sixteen robots, four minutes after the robots have been switched on at the extreme left of the room. Rather than being spread out evenly, they continue to be disproportionately bunched at the far left.

in Figure 11.3. In this case, a number of robots are stuck at the far-left wall. So the best conclusion to draw is that the probability that the robot will move left in one minute is somewhat *lower* than the probability that it will move right in one minute. This is because of the imbalance in the original distribution of robots: the robots were not sprinkled randomly throughout the room, but were bunched up towards the leftmost end, where there was a barrier to prevent them going any further left. This imbalance still remains, and renders the probabilities asymmetric.

Notice that, if we were shown the robots much, much later, when they had had time to disperse evenly throughout the room, then it would indeed be the case that each robot would be equally likely to move left or right in the next minute.

This sort of explanation is similar to the explanation in terms of the low-entropy past. The dynamics of carbon-14 do not inherently favour decay in the future rather than the time-reverse of that process. But given the low-entropy past, it is much more likely that any given carbon-14 atom we see was not spontaneously constituted, because that would involve entropy having decreased. In the very, very distant future, when the cosmos is at, or near thermodynamic equilibrium, we would expect to see carbon-14 being spontaneously constituted just as often as carbon-14 undergoes decay. The chances really would reflect the symmetry in the laws.

But why (you might ask) is it relevant to this explanation to mention the *availability* of the information that the robots were initially distributed at the far left of the room? Surely the chance that the robot will move left in the next minute is less than the chance that the robot will move right in the next minute, because of the initial distribution of robots. Our *knowledge* of this distribution does not affect the chance.

At this point, I need to remind the reader of some of the motivations for adopting a characterisation of chance in terms of available evidence, revisiting some of the material from §4. Most importantly, we noted that alternative characterisations of chance risk *trivialising* all chances. If chance is supposed to be the advice we get from an especially good advice function, it is important to characterise the sort of advice so as to distinguish it from the unhelpful, but sound advice: 'believe all and only the truths!' In this instance, we appeal to the knowledge that the robots started out at the far left of the room, and this informs our estimate of the chance that a robot will move left in the next minute. But we do not appeal, without restriction, to any facts whatever. We do not appeal, for instance, to the fact that this particular robot *does move* one metre to the left in the next minute. If we did that, the chance would be trivial.

So our offering non-trivial answers to questions about chances presupposes that the advice given by the chance function is in some way limited. It is advice given on the basis of evidence that is less than complete. I suggested that the way in which the evidence is incomplete is because chance advice is drawn from evidence that is available, or, in other words, evidence that it is practically possible to obtain.

I also noted that what evidence we count as available may vary in virtue of context. Think about one robot engineer talking to another about the chances that the robots move one way or the other. These engineers, because they are interested in the intrinsic properties of robots, disregard information about where the robots are placed in the room. Hence they formulate chance claims, in the form of dynamical laws, which are temporally symmetric. Much the same happens with nuclear physicists, formulating laws of radioactive decay. The fact that the cosmos started in a particularly low-entropy state is treated as 'unavailable' in this context. A time-symmetric chance law reflects what we know about the way carbon-14 behaves over time, in the absence of any other information. But for us, making our ordinary judgments about what is likely to happen to an atom of carbon-14, versus what is likely to have happened in its past, there is a more expansive notion of what evidence is available. We can draw on the common-sense knowledge we have about the entropy gradient of the universe. So for us, in our ordinary contexts, the chance of decay in the future is different from the chance of spontaneous constitution having occurred in the past.

When I claimed, earlier, that there is much more evidence available regarding the past than there is regarding the future, I noted that this claim is somewhat controversial. Authors like Huw Price (1996: 150–1) have claimed that the asymmetry of traces – whereby we can correlate present events more closely with past events than with future ones – 'fades out' at the level of microscopic systems. So for systems like a single carbon-14 atom, it is doubtful that there are any available traces of its past.

To some extent this is true: if the basis of our knowledge of the past is the past hypothesis, then our knowledge about the past depends upon a fact about entropy. Claims about entropy are claims about large aggregates of particles. Such claims are meaningless or false when applied to microscopic systems involving just a few particles. So appealing to the past hypothesis in the case of a microscopic system does not do the same work that it can do when we look at a macroscopic system. Hence we do not have available, regarding microscopic systems, substantially more evidence about the past than about the future.

But Price, in playing up this point, is neglecting the possibility of shifting context and treating a microscopic system as part of a larger whole – the world. When we do that, we obtain statistical grounds for thinking that elements of the system are likely to have been part of systems that have evolved from lower-entropy states to higher-entropy states. This might not tell us much about whether a given electron was in a spin-up state or a spin-down state one minute ago. But it may well be enough for us to draw conclusions about atoms of carbon-14.

I take Price's examples of temporal symmetry of chances and the like at the microscale, then, to nicely illustrate the very phenomenon of context-sensitivity which is integral to my understanding of chance.

12 | Debunking chance

Chance, when strictly examined, is a mere negative word, and means not any real power which has anywhere a being in nature.

(Hume 1902 [1777])

Darwin's theory of evolution is an unsettling idea. It provides explanations for the existence of many features of the biological world. Not least, it provides explanations for the existence of human traits and behaviours which we typically take for granted. These sorts of explanations can threaten our ordinary self-understanding.

Normative concepts, like rationality, reason, right, and wrong do not play any essential role in the explanations offered to us by evolutionary theory. Moreover, it appears as though the very existence of our concepts of rationality, reason, right, and wrong might be susceptible to being explained – in large part – by evolutionary theory. The question then arises: are the explanations of these concepts that we can obtain from evolutionary theory capable of *vindicating* our ordinary practice in deploying those concepts? Having understood evolutionary theory, should we be content to employ normative concepts in much the same way that we did before? Or does an evolutionary account *debunk* the normative realm? Does it show that, implicit in our normative practices, there is something misguided, erroneous, or otherwise incorrect?

In my opinion, evolutionary theory undermines at least some of our normative concepts. In particular, I am sceptical about our *moral* concepts, in light of an evolutionary explanation of why we have the capacity and propensity to use such concepts. In this chapter I present, in outline, the case for debunking our moral concepts on evolutionary grounds. I then compare moral concepts, and the challenge that they face from evolutionary science, to our concept of *chance*, and the challenge that chance concepts face from a similarly naturalistic account of why we have acquired them. The upshot of this comparison is that, although our concept of chance and its associated norms may be very *useful*, many of our ideas about chance seem to be debunked by an evolutionary explanation. This supports an anti-realist attitude towards chance.

51 Norms and vindication

Take the concept of *moral wrongness*. This is a central and typical example of a moral concept that we use. We use it, for instance, in the way that we condemn many sorts of violence. What does the concept of moral wrongness involve? There seem to be two or three key features of the concept.

First, and most obviously, we understand that if an action is morally wrong, we have some sort of reason not to do it. So the concept is *action-guiding*. Judgments that something is wrong contribute to determining what we ought to do.

Second, we typically take the action-guiding role of moral wrongness to be *universal*. Everyone has the same sort of reason – in virtue of some behaviours being wrong – to abstain from the prohibited behaviour.

This second thought requires some qualification. There might be some – such as the very young or the mentally incompetent – who we think are not properly guided by moral judgments like this. Such individuals, we think, are not capable of seeing the force of a moral judgment, or should not be held to the same standards as us, or some such. But putting such cases aside and focusing on those who have much the same psychological apparatus as we do, we think that judgments of moral wrongness have the very same force for all of us. Diversity of taste, whim, or idiosyncrasy cannot in themselves explain or justify ignoring that something is wrong. I cannot say that, 'I understand very well that killing is wrong, but I don't really go in for behaving morally.' To speak like this is to suggest that I do not understand what a moral judgment is.

Third, and this is in many ways just an elaboration of the previous two features, the action-guiding force of moral wrongness is in some sense *inescapable*. This is an important contrast between moral norms and other normative concepts. Take, for instance, legal norms. For most of us, most of the time, the law has normative force: we take the law to provide us with reasons to engage in and to avoid certain behaviours. Now imagine a case where we might be seriously tempted to break the law. You might be asked, by your terminally ill relative, with very poor prospects for future quality of life, to assist her to end her life. You might want to run a red light in a small country town, where there is no danger of an accident, in order to get to an interview on time. You might ignore a planning law, because of your long-cherished ambition to have a house with a cupola, despite its inconsistency with the heritage values of the site.

In such cases, the agent in question is no longer taking the law to have normative force. The agent is declining to 'play by the rules'. But it does not seem that one's behaviour as an agent, or as a norm-follower in general, is seriously defective if one decides to ignore the law in such a case. It simply seems that one is responding to other – arguably more pressing – reasons.

Outside of the legal context, there are other, even less controversial examples. I might wish to ignore the rules of etiquette, in order to enjoy myself more at a party, where everyone else is ignoring the rules of etiquette. I might wish, in order to play a foolish joke on my opponent, to ignore the rules of tennis by lifting up the net at a decisive moment in a rally. These behaviours might make me a poor follower of etiquette, or a poor tennis player, but they do not necessarily make me irrational, or show that I have trouble grasping the force of norms in general. Rather, they show that the norms of etiquette, of tennis, and even of the law, are in some cases 'escapable'. It is conceivable that I could be in a circumstance where I can reasonably choose to defy those norms. This looks like a potential difference between the norms of morality and other sorts of norms. Moral norms are either outright inescapable, or they are at least more difficult to escape than most other sorts of norm.

So these are our three characteristics of moral judgments. They are:

1. action-guiding,
2. universal, and
3. inescapable, or near-inescapable.

Vindicating non-moral norms

How could we 'vindicate' a set of normative concepts? Let's begin with the non-moral norms – the easier cases.[1]

One way would be to argue that there is a substantial benefit in adopting the system of norms. This would be a sort of instrumental or *practical vindication*. The idea here is that if you show that people do better with the norms than without, then you have shown that people have reason to use or deploy those norms.

This sort of vindication, if it works, will identify non-trivial *hypothetical imperatives* of the form: 'If you want X, then (you ought to) do Y.' So we might vindicate conditionals such as:

---

1 My approach here is inspired by an influential article by Philippa Foot (1972). Foot, however, is interested in undermining the traditional idea – which I am endorsing – that morality is distinguished by consisting of categorical imperatives.

- If you want to prosper at a boys' school in England, you ought to obey the rules of cricket.
- If you want to avoid legal penalty, you ought to obey the law.
- If you want to be a welcome guest at parties, you ought to adhere to a similar level of etiquette as the other guests.

Of course, not just any hypothetical imperative will suffice to 'vindicate' a set of norms. If we think that the goal X which is served by following the norms is unappealing, or simply less impressive than we might have hoped for, then there is something *deflating* about such a vindication. For instance, having thought about the benefits to be obtained by following the law, in a Marxist vein, one might suggest that:

- If you wish to endorse norms through which the bourgeoisie oppresses the working class, and to subscribe to a false consciousness, then you ought to obey the law.

Such a dismal hypothetical is not much of a vindication of legal norms.

A second type of vindication would be an attempt to demonstrate the existence of the norms: call it *existential vindication*. To do this, we take particular judgments in that system of norms and attempt to understand the conditions in which those judgments are true. Take, for instance,

- It is a rule of cricket that a player will be judged leg before wicket in the following circumstances . . .
- It is a norm of etiquette (around here) that it is bad manners to burp while one is eating in the company of others.
- It is a law of Victoria that theft is punishable in accordance with section 74 of the Victorian Crimes Act (1958).

I take it that the truth of these sentences requires that there be a community of individuals who engage in a set of practices that somehow 'support' the norm. H. L. A. Hart's *The Concept of Law* (1961) stands out as a landmark attempt to identify the sort of social behaviour required for legal norms to exist. Other norms, such as those of etiquette, or the rules of a sport, can be understood as somewhat less regimented variations on similar ideas.

To existentially vindicate the norms would then require that we confirm that these very conditions obtain. This is perhaps less philosophically interesting, but there is an important point to note about vindicating these norms, in the existential sense. In many cases, it is very plausible that your investigation of the norms will bottom out in cases where there is *indeterminacy*. Among reasonably well-informed people, once you secure agreement on the social practices, you will secure widespread agreement on what the norms are. But you will find that some questions of etiquette are simply not settled by the social practices. Occasionally, we engage in argument about these things beyond what the social practices will bear. Good examples are common in the norms of correct English. Is it correct English to use 'disinterested' and 'uninterested' interchangeably? Does 'I could care less' now mean the same as 'I could not care less', or is it that large parts of the population say something false when they utter the former? These are questions of conformity to norms where the norm has changed, or is arguably undergoing change, and although we may rather enjoy engaging in arguments over these matters, in more reflective moods we admit that we should be open to the possibility that there is no correct answer.

It is more controversial whether this will happen with law. Ronald Dworkin (1977, 1986) has argued, contra Hart, that there will always remain existential questions about law that remain *properly disputable*, without being capable of being settled by appeal to social facts. So Dworkin's understanding of law is that it is not settled by social fact alone. But even if we concede to Dworkin that there are some persistently normative disputes in law, over and above disputes about the social facts of the relevant society; even if we allow that it is legitimate to engage in persistent dispute over the correctness of phrases such as 'I could care less' versus its negation; we can still agree that *large swathes* of these systems of norms are *settled by social practice*.

Vindicating moral norms

Why is it any different to vindicate a moral norm?

First, regarding practical vindication, a crucial difference is that moral norms are supposed to give rise to non-hypothetical demands. Morality gives rise to 'categorical imperatives'. So the sorts of reasons to which moral norms give rise are presented to us as obligatory. We must respect those reasons, regardless of our particular ends and aims. This follows from the inescapability of moral norms.

As such, if we are going to attempt a practical vindication of moral norms, in a manner like that described for non-moral norms, then it will prove to be an inadequate vindication. In much the same way that the Marxist hypothetical imperative is 'deflating' – it ends up vindicating the system of norms for the wrong reasons – it seems that *any* prudential or pragmatic justification of moral norms is going to be similarly deflating. It is insufficient to justify what we intended to justify.

Second, if you try to engage in the more anthropological, merely existential vindication of a moral norm, that proves problematic too. You find that – compared to the less controversial norms – it is much harder to get agreement on the existence of moral norms, even when participants to a moral debate agree entirely on the social practices that exist. This is in part because of the *universality* of moral norms. Moral norms do not generally present themselves as merely local customs, such as driving on the left-hand side of the road. Rather, they present themselves as directives which should govern those of us who reject them just as much as those who currently accept them.

So, for instance, it is very hard to find uncontroversial truth conditions for sentences like:

(1) It is a universal moral rule that infanticide is always wrong.

Many well-informed people will insist that sentences like this are true, even though we may point out the lack of social support for such a rule in various places and at various times. Others, similarly apprised of the facts, will insist that the contrary is true: that infanticide is not wrong under certain circumstances.

If we were less ambitious, we could seek existential vindication of more local rules. For instance, I imagine we can identify uncontroversial truth conditions for:

(2) It is a widely accepted rule in community *C* that infanticide is always wrong.

And this is consistent with finding that the rule is universal in scope. That is, we can find that the members of community *C* take their rule to govern not just their own behaviour, but everyone's. But this is not the sort of existential vindication we are after, either. Non-moral norms typically do better because non-moral norms do not present themselves as universal and inescapable. In the case of cricket, we can vindicate the existence of rules by looking at the social practice in cricket matches. There is a pleasing correspondence between *what* social behaviour the rules are supposed to govern and *where* the relevant social basis is. If we find a community of individuals who seem to reject the rules, then we can simply conclude that they are not playing cricket. But moral rules, understood as universal and inescapable, are underdetermined by any social facts in the domain they are supposed to govern. At best we get either non-universal rules or an abundance of universal rules which contradict one another.[2]

A further, final point can be made about the difficulty of providing an existential vindication of moral norms. Even if there were a universal practice of condemning infanticide, it does not seem incoherent for someone to ask: but is infanticide *really* morally wrong?[3] It seems, then, that moral norms are at least not a priori reducible to any pattern of social fact.

This is a puzzle. Why are we so confident that moral norms *do indeed exist*? What is it that we are so confident of?

52 The natural history of moral norms

Having run into this sort of difficulty, an evolutionary account of moral norms promises to explain why we might have acquired this system of norms that is so difficult to vindicate. Below I give a quick sketch of some recent work on the evolution of morality, drawing largely on the work of Richard Joyce (2006).[4]

First, we need to recognise that with our current psychological make-up, there is a tendency to be tempted towards short-term gain, at long-term cost. We are good at rationalising decisions to ourselves, and this can lead to our talking ourselves out of what we earlier judged would be best. This is typical in instances of weakness of will. Imagine yourself, on a Sunday morning, drinking some coffee and reading the newspaper. You have planned to get started on some home repairs today. To do so, however, will require stopping your breakfast earlier than you would like, making some measurements, and

2 Having stressed the existence of widespread moral disagreement, it should be noted that there is some universality in terms of what sorts of behaviour are apt to be regulated by moral norms. Reciprocal relations, violence, status, and certain matters relating to bodies and bodily functions are all subject to moral norms in different cultures; but the exact content of those norms varies greatly. See Haidt and Joseph (2004) for a psychological model that attempts to explain the commonality and diversity of moral judgments across human societies.
3 This is a variation on G. E. Moore's 'Open Question' argument (Moore 1993).
4 Note that, in putting forward this 'evolutionary' explanation as part of a debunking argument regarding morality, it is not essential – for the success of the argument – that the mechanism by which we acquire moral norms be natural selection. All that is required is that we give a genealogical account, which explains *why we have* these norms, without presupposing that the norms are tracking something true or that the norms have some independent justification. As it happens, the genealogy of moral norms does indeed seem to involve natural selection. But when I go on to discuss chance, it will turn out that natural selection plays a less significant role in the acquisition of the distinctive conceptual features of chance. Given this, I prefer to use the term 'natural history', rather than 'evolutionary account', to make clear that the particular mechanism need not be natural selection, or even any other form of evolution.

driving down to the hardware shop. Although in a cool moment you judged that it would be worth curtailing breakfast in order to start this job, you are strongly tempted now to revise that judgment. You begin to rationalise to yourself, and soon you have come up with a justification for putting off the repairs until next weekend: 'I had a hard day yesterday. I deserve this.' Or, 'It looks like it might rain, anyway, so it will be a bad day for working on the house.' Consequently, you abandon the earlier intention.[5]

Your ability to engage in rational deliberation is, of course, a huge benefit. It enables you to plan and execute complex, lengthy projects, like planting and maintaining a vegetable garden, building a house, writing a thesis, or migrating to another country. But it appears to make us vulnerable to a distinctive form of procrastination that is encouraged or reinforced by our ability to rationalise our decisions to ourselves in the face of temptation. The vulnerability to this sort of weakness of will, for instance, certainly does not seem to be a part of the psychology of less cognitively sophisticated species. Imagine a beaver, for instance, undertaking some repair work on its dam. Not only do beavers not appear to be subject to weakness of will, it is scarcely conceivable that they could both be weak-willed *and* be so with the aid of *rationalisation.*

Think of this vulnerability associated with rational deliberation as the psychological background condition in which a capacity for moral thought and judgment might turn out to be adaptive (Joyce 2006: §4.2).

Second, it is especially important, for the success of cooperative endeavours with other humans, that we be able to engage in long-term commitment to collective goals, and in the process abstain from seeking short-term individual goods (Joyce 2006: §4.3). In operating a business with a partner there may be innumerable opportunities to take the assets of the business and run off to a foreign land for a wild time of gambling and debauchery. To some, this might be very tempting. But the benefits of a business partnership will never be open to you at all unless you can gain the trust of your partner that you will never succumb to such temptation. Given that the benefits of collaborative endeavours with other humans are enormous, overcoming the tendency to short-term gain and rationalisation is potentially a very beneficial adaptation.

Third, if some agents evolve a heritable mechanism which enables them to effectively eliminate from deliberation trust-breaking options such as running off with the business assets, then this could be an adaptive

---

5 For an instructive discussion of weakness of will and its distinction from *akrasia* – acting against one's better judgment – see Holton (2009: chap. 4).

mechanism for two reasons. First, it may have direct prudential benefits, because it may enable one to participate more often in productive collaboration with others. There may even be benefits for personal enterprises that do not involve collaboration with others, but require the fixed pursuit of a goal for a long period of time. Second – though this claim is much more controversial – it may have group selection effects, if groups with this trait are able to out-compete other groups without it (Joyce 2006: §1.6).

The core suggestion, then, is that the ability to make moral judgments, and to develop sets of moral norms, *is* a heritable mechanism in humans which has enabled us to obtain the fitness advantage of reliably entering into trusting collaborative arrangements.[6]

Importantly, this account partly explains why there should be a distinctively unconditional character to these norms. Because moral reasoning has evolved to overcome an existing design weakness – namely, that our rationalisation abilities are so good that we can talk ourselves into virtually anything – we need the moral judgments to have a certain *inescapability* if they are to be effective.

We thus get an explanation of *why we have* a system of categorical imperatives *without* getting a practical vindication. At best we get a deflated vindication: it turns out that, for many of us much of the time, being moral is useful. Or, more precisely, it turns out that for many of our ancestors, most of the time, the capacity for moral judgments was useful. And, moreover, that utility is not necessarily the sort of utility that we care about – utility for *us*. Rather, it is utility for our *genes*: an advance to our fitness.

As to existential vindication, the explanation that we have given of the capacity to form moral beliefs did not require that there be any mind-independent facts to which those beliefs correspond. So we have no reason to believe that our moral beliefs represent facts. Moreover, the evolutionary hypothesis sketched here suggests an explanation of why we *ordinarily think* of moral facts as mind-independent, universal, and objective. If we thought of our moral beliefs as mere subjective attitudes or dispositions, then they would likely be less effective in combating our tendency to rationalise.

So, at best, an evolutionary account gives us a deflated practical vindication of moral norms, and throws grave doubt on the existential vindication. Moreover, it gives us reason to be pessimistic that any better vindication

---

6 The account I have given in the text is of necessity hugely abbreviated. In particular, I've omitted to mention the very important role that punishment seems to play in establishing the mechanism of norms. See Joyce (2006: especially §§1.4, 2.6).

will be found. If the evolutionary explanation works, any further attempted story about moral facts will appear redundant.

53 Chance compared to morals

Our concept of a chance, like our concept of a moral norm, has an important normative element. The central normative idea regarding chance is that we should match our degrees of confidence – our credences, to use the preferred lingo – to what we believe the chances are.

The classic example of someone who violates such a norm is a gambler, convinced that the roulette wheel is properly constructed, convinced that there is no probabilistic relation between past spins and future spins, but nonetheless sure that the next spin will come up on a particular number. The gambler believes the *chance* of that number coming up is $\frac{1}{37}$, but he is nonetheless extremely *confident* that this number will come up. His confidence manifests in his strong tendency to place bets on that number to the exclusion of all others, whereas his belief is manifested in his conscious thought about how the wheel is constructed, how the symmetry of the wheel corresponds to a symmetry of chances, and so forth. This is a rational defect: the gambler is defying the requirement to match his beliefs about the chances to his degrees of confidence.

But this first norm is not enough. What about agents who have dubious beliefs about the chances? If the gambler is convinced that the chance of getting 17 on the next spin of the wheel is in fact $\frac{1}{3}$, then we might think the gambler will be more successful if he flouts the imperative to match his degree of confidence to his beliefs in the chances. To address this concern it becomes necessary to impose some further constraints on how agents respond to evidence, and thereby revise their beliefs in the chances. I suggest that something like the imperative: 'Adopt beliefs in the chances that, other things being equal, maximise the likelihood of your evidence', is the relevant norm here. We needn't worry, for our purposes, about a more precise specification.

Because we will be referring back to these ideas, it will be useful to have names for them. Something very like the first norm is widely known as the 'Principal Principle', but I shall call it *coordination*, so as to avoid any unnecessary connotations with the rather intricate recent debates over the precise form that the principle should take.[7] All we require is the basic

---

7 See the discussion above at §4, and references therein.

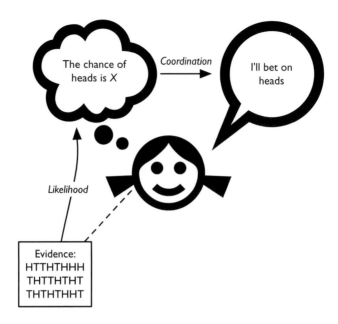

Figure 12.1 The two norms that are involved in chance belief. A rational agent will adopt chance hypotheses that make the evidence more likely (likelihood); and will also use those chance beliefs to regulate her degrees of confidence, and hence her behaviour (coordination).

idea that beliefs about chance should match one's degrees of confidence, where those degrees of confidence are understood to have some direct link to behavioural dispositions.

The second norm I will call *likelihood*. To reiterate, it is simply the idea that one should adopt chance beliefs that make one's evidence more likely. Again, more precise versions of this idea have been proposed and debated, and I wish to abstract from such details. The schematic relationship between the norms is illustrated in Figure 12.1.

I take coordination and likelihood to be the two most important norms regarding chance. In addition to these norms, however, there are other conceptual features of chance that need to be identified in order to distinguish chance from other sorts of probability. Although there is no definitive list, the following two features seem to me to be essential:

Objectivity What the chances are does not depend upon the particular evidence one has: it depends on something reasonably universal, so that believers in even quite different epistemic circumstances are trying to match their credences to the same chances.

It is important to note this, because the second norm regarding chance refers only to the evidence of an individual. But there is a presupposition about chance that it is universal and the same for all of us, at a given time. (See Chapter 1 for further discussion of the way in which chance is objective.)

Irreducibility Chances cannot be reduced to frequencies, or any other purely 'actual', non-chancy facts. Consequently, matching your credence to the chances is no guarantee of success. The past evidence regarding chances can be grossly misleading, and the future frequencies of events may fail to match the chances. (See Chapter 7.)

Finally, a fifth idea is perhaps not essential, but is certainly part of the understanding of chance which many people now have, in light of the idea that chances could feature in our fundamental physical theories of the world.

Physicality Chances are in some sense physical properties of the world. This has at least two important connotations: (i) Chances are grounded in the intrinsic properties of physical phenomena. So experimental set-ups which are intrinsically similar will exhibit the same chances. And (ii), chances play a role in explaining physical phenomena much like the role that non-chancy causal factors play in explaining physical phenomena.

So, to summarise, we have five key ideas regarding chance that define the target of our attempted vindication: (1) a normative chance–credence link (coordination); (2) a second normative requirement to adopt chance-hypotheses that increase the likelihood of our evidence (likelihood); (3) the objectivity of chance; (4) the irreducibility of chance; and (5) the physicality of chance.[8]

Initial troubles with vindication

To vindicate the norms of coordination and likelihood, in a practical sense, we would need to identify a suitably non-trivial hypothetical imperative of the form:

If you want to X, then you ought to follow the chance norms.

First, the good news. This does not look like such a hopeless task as it did with the moral norms. We can probably come up with a very general goal, such as having a high number of true beliefs about the outcomes of chance events, or being successful in trying to achieve one's goals, which, if it is

8 Of these five ideas, the one that is most obviously controversial is irreducibility. It would be rejected, for instance, by frequentists, or by David Lewis, who advocated a very sophisticated form of reducibility of chance to frequencies. But these views are driven by other philosophical motivations, and are generally recognised as departures from the naive concept of chance. So I propose to begin by examining whether the chance norms can be vindicated, while retaining this principle. Of course it remains open that a variation on our chance concept, without this feature, will fare differently.

indeed served by following the chance norms, would be a non-deflating practical vindication of those norms.

So in this spirit, let's propose that we can practically vindicate chance if we can justify the following:

(3) If you want to achieve your goals, then you ought to follow the chance norms.

This seems much more achievable than trying to vindicate a categorical imperative.

Now for the less good news. In trying to provide a rationale that justifies (3), it is extremely hard not to end up presupposing something about chance and the chance norms. Consequently, it looks as though we will at best achieve a somewhat circular vindication. To see this, reflect on the fact that a conditional like (3) does not describe a surefire recipe for success. It is only *probably true* that, if you match your credences to the chances, you will be successful. There may be people who are unsuccessful, despite matching their credences to their beliefs about the chances, and despite adopting beliefs about chances that maximise the likelihood of the evidence.

Admittedly, much the same sort of thing could have been said about our attempt to vindicate the non-moral norms in a practical sense. It is not true for *everyone* that, if they want to be a popular guest at parties, they should match their level of etiquette to that of the other guests. For some, it will be a better strategy to be deliberately outrageous, and to flout the rules of etiquette. Others, despite their best efforts to conform, will remain unpopular. So the hypothetical imperatives we cited before are surely only rules of thumb, rather than promises of surefire connections between compliance and success. So perhaps it is too demanding to ask for a 'surefire' hypothetical imperative. On the other hand, if we do not demand a conditional that gives a strict formula for success, there is a clear danger of circularity. In justifying the imperative, we do not want to be presupposing:

(3*) There is *a high chance* that you will succeed in achieving your goals, if you comply with the chance norms.

We cannot invoke this sort of claim in the process of vindicating the chance norms, because it is only relevant if we can presuppose the very norms which we are trying to vindicate!

Circularity of this sort might be unavoidable. I suspect that probabilistic notions infect so much of our thinking that a vindication of chance norms will have to use at least some probabilistic ideas. I do not think that is necessarily fatal: the circularity could be a virtuous one. But I find it very hard to be confident of my opinions in this matter. So I suggest that we put this problem aside.

What about existential vindication? Here we run into a problem that is extremely similar to that which beset the existential vindication of morals. Recall that, even where people agree on the social practices in place, there can be disagreement about what the moral norms really are. It may be uncontroversial what the norms are *believed to be* in a particular community. But because the norms are believed to be universal, it is not possible to identify such merely local norms as genuine *moral* norms.

Similarly with chances, even if we are agreed on the frequencies, this does not settle what the chances themselves are, because of the irreducibility of chance. Moreover, it turns out that the second chance norm – likelihood – is itself quite permissive. There are many ways of responding to the evidence, so as to adopt chance hypotheses which make the evidence more likely. So even if we reduced our ambition, and sought agreement only on what we should all *believe* about the chances, given a large pool of shared evidence, that goal may still be unachievable.

So the debate is getting difficult here, and we are not finding any new purchase on it. I do not claim to have shown that the vindication of chance has unequivocally failed. Moreover, there is reason to think that practical vindication will be easier for chance than it is for morals. But, that said, it is hard to get an independent handle on why chance norms are practically beneficial, because any attempt to state precisely the way in which they benefit us tends to mention chance. Regarding existential vindication, we run into a problem directly analogous to the problem for morals: the irreducibility of chance is akin to the irreducibility of moral rules to social practices.

In order to move forward, then, I suggest we turn again to a natural history perspective. This might help us to see, in non-chancy terms, what benefit is served by having the cognitive apparatus which permits us to entertain chance beliefs and to comply with the chance norms.

54 The natural history of chance

The psychology of chance

What does it take to be an agent who has chance beliefs? What epistemic constraints must such an agent observe in order to be rational?

I do not have anything like an adequate *empirical* answer to this question. I do not know, in humans, what it is that enables us to entertain thoughts about chance. But there is a strikingly effective model of the epistemology of chance which has come to wield considerable *normative* force in philosophical discussions. This is the framework of Bayesian epistemology, and in particular the idea of *exchangeability* developed by de Finetti (1937). This model, I must stress, is not particularly well supported as a model of how we actually think. But as a sparse, formal approach to what sorts of constraints an agent must respect in order to be rationally thinking about chance, it has met with wide support.[9]

I discussed these ideas above, in some detail, in §37, but for the reader who has jumped into the discussion more recently, here is a brief outline of the framework.

Probabilistic beliefs

Agents need to have views about probability in order to choose how to act. Intuitively, agents weigh up the likely outcomes of different actions, and the relative value or disvalue of those outcomes, and this gives some basis for choosing one act or another. For instance, in deciding whether to purchase a ticket to Sydney flying on Qantas or another airline, I might have considered the possible outcome that my flight might crash, leading to my injury or death. Taking that to be a relatively bad outcome, I wanted to choose an airline that minimised its likelihood.

Bayesian decision theory gets at this probabilistic information, implicit in our decisions, by beginning with our preferences for different actions. Suppose that, by some means or other, we come to rank available acts in a preference ranking. This psychological state suffices to determine two functions: a probability function for the possible outcomes, conditional on each act, and a utility function for those outcomes. An agent who chooses the most preferred act out of all those available to her will be behaving as an agent who maximises expected utility according to these implicit functions.

So an agent who has the right sort of structure of preferences among actions, *ipso facto*, is an agent with probabilistic beliefs about the outcomes of

---

9 In my own case, I do not even claim that this approach is the best of those currently being debated. I am not sufficiently familiar with the alternatives to give a useful judgment. I simply present the Bayesian account here because it is, effectively, the default framework in current discussions.

actions. Note, that 'outcome' is to be understood very broadly here – we don't just mean causal consequences of actions, but also non-causal outcomes such as *It turned out that I was right to say 'England will win'*. Consequently, the sort of probabilistic beliefs that agents have will be sufficient to cover any proposition about which the agent has an opinion at all.

Changing beliefs over time

Our probabilistic beliefs, however, are liable to change. And, correspondingly, our preferred acts are liable to change. How should a Bayesian agent respond to new evidence? It is possible to analyse an agent's attitude towards new evidence in the same preference-based framework as before. An agent may have preferences to choose later acts *on the basis of the outcomes of earlier acts*. For instance, offered two unknown flavours of ice-cream, I may think the probability I will like flavour A is 0.8, and the probability that I will like flavour B is 0.9. Correspondingly, I choose flavour B.

What is my preferred outcome for future choices, conditional on the outcome of earlier trying flavour B? If, on the first taste, I like flavour B, I prefer even more strongly to choose flavour B in future. If, on the first taste, I dislike flavour B, I prefer to try flavour A next time. These conditional preferences amount to *conditional probabilistic beliefs* about whether I will enjoy flavour B the second time I try it, given how much I liked it the first time.

Suppose I always revise my preferences so as to conform with my conditional preferences, once the outcome of an action is revealed to me. Then it will turn out that my probabilistic beliefs – as determined by my preference ranking – will have undergone revision in the way known as 'updating by conditionalisation'. This is simply a more formal and precise way of formulating the norm of likelihood: the requirement that one adopts chance hypotheses which make one's evidence more likely.

Chance beliefs

The framework developed so far is dealing with probabilistic beliefs in general, as opposed to belief about physical chances in particular. I take it that a chance is a special type of probability, and that only some of our probabilistic beliefs are plausibly thought to be beliefs about chances. To be a belief about chance, we typically expect the person who holds such beliefs to have associated ideas such as those I identified above: objectivity, irreducibility, and the two norms of coordination and likelihood.

The Bayesian understanding of chance psychology approaches these aspects indirectly, by focusing on the distinctive attitude we have to evidence that bears on a matter of chance.

Paradigmatically, we conceive of instances of chance processes – such as coin tosses – as independent, identically distributed trials. The outcome of the 99th coin toss has the same probability of heads as the 9th coin toss and the 999th toss. There is no probabilistic dependence between trials. The coin is no more or less likely to land heads on the tenth trial, having landed heads on all the previous trials, than it would be if it landed tails on all the previous trials. Believing that the coin tosses are identical, independently distributed trials is symptomatic of believing that the coin-tosses are matters of chance.

Drawing upon this sort of case as the paradigm of a chance process, de Finetti has shown how we can characterise, purely in terms of preferences over possible acts, the psychology of an agent who is entertaining a number of distinct chance hypotheses.[10] The key feature here is that one treats the trials as *exchangeable*. If I find out that you tossed three heads first followed by seven tails, I treat this evidence in much the same way as if I found out that you tossed the same numbers of heads and tails (three and seven), but more evenly interspersed. The *sequence* in which the data arrives, or is collected, is irrelevant.[11] But if I was trying to calculate the probability that my child would have a fever today, on the basis of the past ten days, there would be an enormous difference in my subjective probability, depending upon whether my child has had three days of fever followed by no fever for seven consecutive days, versus the case where my child has had three days of fever scattered through the past ten days. I do not treat each day's temperature as exchangeable with other days: the order matters to my epistemic state. So I do not treat my child's fever on any given day as a matter of chance. (Each day is not an independent trial for having a fever: the probability depends, in part, on whether or not one had a fever the previous day.)

Note that it is hard to say what makes it the case that a probabilistic phenomenon in the world really does involve independent, identically distributed trials. If we could do that we would, in effect, have resolved the

---

10 See de Finetti (1937: 118–30) for the notion of exchangeability. For a somewhat more accessible presentation, see Skyrms (1984: 20–36).

11 We do not, in practice, regard any empirical trials as *strictly* exchangeable. For instance, if I tossed a coin one thousand times, and obtained 500 consecutive heads, followed by 500 consecutive tails, you would not think that this confirmed the hypothesis that the coin is fair. Rather, you would be pretty near convinced that I have cleverly rigged the coin. So the requirement of exchangeability should be understood as an ideal – the way an agent would behave if she were a priori committed to the chanciness of a given phenomenon.

existential vindication of chance. But, as I argued above, this is difficult because of irreducibility. More modestly, de Finetti hopes to have given us a precise formal characterisation of *what it is for an agent to treat a phenomenon as a chance process*. No story has been provided for when that is an appropriate or correct attitude to adopt.

So, in effect, the Bayesian approach to understanding chance psychology suggests that an agent is behaving as if a phenomenon is a matter of chance if, and only if, she meets the following requirements:

- she has a sufficiently rich set of preferences over possible acts;
- those preferences have the right structure (they are transitive and constitute a total ordering);
- she revises her preferences for acts in accordance with her conditional preferences; and
- her conditional preferences are such that she treats repeat trials of this phenomenon as *exchangeable*.

Further, in order to capture a suitable sort of open-mindedness, we need a non-dogmatism requirement:

- She is *open* to all possible chance hypotheses. (The agent does not rule out any hypotheses as a priori impossible, though she may give them extremely low probability.[12])

Because of the requirement that the agent revises her preferences in accordance with her conditional preferences, an agent complying with these requirements will be complying with likelihood. Moreover, she will in some sense be complying with coordination, because her beliefs about the chances will simply be implicit in her degrees of confidence, which are in turn grounded in her structure of preferences.

It is not clear, however, that an agent who meets these requirements will be a believer in *objective* chance hypotheses. Nor is it clear that an agent meeting these requirements will be believing in chances that are *irreducible* to frequencies or other empirical facts. But defenders of the Bayesian approach have cleverly shown that there is something very like objectivity implicit in the chance beliefs of an agent meeting these requirements. That is, different agents – provided they regard the same trials as exchangeable – will behave

---

12 Expressing this requirement mathematically is a bit tricky, given that, if there are infinitely many possible chance hypotheses, it cannot be that an agent assigns a positive credence to all of them while maintaining coherence. But we need not worry about these technical matters here.

as if there is a common optimal strategy, and will defer to the opinion of more informed agents who know more about the outcomes of such trials.

> The optimal strategy is not subjective, in the sense of being the betting strategy of any agent. It is something that [a Bayesian agent] regards as optimal for bets on a certain class of experimental set-ups. Furthermore, where he conditionalises on the results of elements of the sequence, he learns about what the optimal strategy is, and he is certain that any agent with non-dogmatic priors on which the sequence of experiments is exchangeable will converge to the same optimal strategy. If this is not the same thing as believing that there are objective chances, then it is something that serves the same purpose.   (Greaves and Myrvold 2010: 282)

Similarly, agents meeting these requirements will behave in a way that is highly suggestive of irreducibility, in virtue of the non-dogmatism requirement and the way in which the mathematics of conditionalisation works. Because of non-dogmatism, a Bayesian agent will assign some non-zero degree of belief to all chance hypotheses. Given the way chance beliefs evolve under conditionalisation, these degrees of belief will never go to zero. So one will never be in a position to rule a chance hypothesis out as impossible. Having tossed a coin a million times and only seen heads, it is still not impossible that the coin is fair, and a Bayesian agent will have credences that reflect this. This captures much of the idea that frequencies can be very misleading evidence of chances, and that one cannot straightforwardly identify chance facts with frequencies, which is essentially the same idea as what I called irreducibility.

One may quibble that these attempts to capture the ideas of irreducibility and objectivity are not quite right. In particular, these ideas might be thought to have metaphysical connotations which have – in the Bayesian approximation – been lost. Moreover, this account does not do anything directly to address the idea of physicality. It is perfectly open to an agent to adopt a chance hypothesis with regard to trials that are intrinsically highly dissimilar. The Bayesian account has nothing to say about why this would be inappropriate. I will return to these concerns below, but for now let's grant the Bayesian that she has adequately explained the psychology of chance. Does that help us, either to *practically* vindicate the chance norms or to *existentially* vindicate particular chance hypotheses?

First, consider existential vindication. Note the different forms this problem could take. We could try to vindicate either of the following:

- Jones is confident to degree x that the chance of heads is p.
- The chance of heads is p.

To vindicate the first, we have enough. The psychological account has given us a clear description of the sorts of conditions that must be met by an agent to have a chance hypothesis of this sort. To vindicate the second, however, we do not yet have enough. The psychological account has done nothing further to tell us under what conditions it is *correct* to adopt a particular chance belief. And even agents who agree on all the evidence and update their beliefs 'correctly' may still disagree, if they have sufficiently different initial credences. So it has not advanced us any further to existentially vindicating chance claims.

What about the practical vindication? Here some progress appears to have been made, but at the cost of it being a *trivial* practical vindication. If you want to meet the requirements of rationality, as stipulated in the above conditions, then you must, *ipso facto*, meet the coordination and likelihood norms. But this does not entail anything about your *success*. Either success is defined internally as revising your preferences to match your earlier conditional preferences – but that's trivial – or it is defined externally, in terms of achieving your goals. Nothing has been said to explain why revising your preferences in this way will help you to achieve your goals. To do that, a substantive story must be told about the link between your psychological states and the world.

Some empirical observations and speculations

In the preceding section I described a commonly used formal model for understanding rational psychology regarding probabilities and chances in particular. Although I cannot hope to give a general account of how we actually engage in cognition about probabilities, there are some empirical discoveries about our cognitive abilities in this area that are worth bearing in mind.

First, many *non-human* animals engage in behaviour that resembles adopting a chance hypothesis. I do not know if there is evidence that animals treat the data as *exchangeable*. But non-human animals, having learned the probability of success of two different strategies A and B, do *maximise* by employing only the higher probability strategy. For instance, in a scenario where operating lever A leads to a reward with 80 per cent probability, and operating lever B leads to a reward with only 20 per cent probability, non-human animals such as pigeons and rats will soon adopt the behaviour of operating lever A exclusively.

If a rational human were convinced that success were a matter of chance, and that these were the probabilities, this is the sort of behaviour the human would engage in too.

Second, many humans are *outperformed* by rats and pigeons in tasks like this! A human subject, initially ignorant of the chanciness in the set-up and given the same levers A and B, will tend to operate lever B about 20 per cent of the time and lever A about 80 per cent of the time. That is, the frequency with which human subjects operate each lever tends to match the probability of success on each lever. And they persist in doing this, even in strikingly long sequences of trials.[13] Human subjects tend to seek a *pattern* to explain the past successes they have had, and this leads them to non-optimal behaviour. Being random, there is no pattern to be identified, so they would be better off simply pursuing the higher probability option.[14]

Clearly, the human subjects in this sort of experiment are *not* treating the sequences of data as exchangeable. They are looking for probabilistic dependencies between trials, and thus treat different sequences of lever pullings and successes as having a different epistemic bearing on what they should do. I take it that this is a sort of *causal reasoning* that humans are engaged in. And, of course, causal reasoning is often hugely rewarding. That sort of reasoning is what enables us to engage in and succeed at all sorts of sophisticated behaviours, such as building fires, growing crops, raising animals, and playing poker. This type of reasoning is part of why we outperform rats and pigeons at so many cognitive tasks. But in an artificially simple case such as this, in which the two levers have fixed chances of success, we reveal a potential *cost* of our cognitive sophistication.

Indeed, other studies have suggested that our persistent seeking for causal explanations of features of the world positively blinds us to some ways in which the world is chancy or random. This seems to be a large part of the reason why we entertain superstitious beliefs. To take a favourite example, Langer and Roth (1975) asked Yale undergraduates to predict the results of thirty coin tosses. The results were manipulated to ensure that everyone achieved precisely 50 per cent success rates. But some results were manipulated so that students had early 'streaks' of success.

Among all the students, *one quarter* reported that they thought their performance would be hampered by distraction. *Forty per cent* said their performance would improve with practice. Moreover, those who had early

---

13 In one study probability-matching behaviour persisted in many human subjects for as long as the experiment ran: over 400 trials (Yellott 1969), though in this case the probabilities and the reinforcement schedule were varied after each block of 50 trials, which might have contributed to the subjects' being slow to adopt the superior maximising strategy.

14 See Wolford, Miller, and Gazzaniga (2000) for a brief discussion of the comparative behaviour of human and non-human subjects in these experiments, and the results of an attempt to identify neural correlates of the distinctive human behaviour.

streaks of success rated themselves as better at 'coin guessing' than the others, even though their overall success rate was – like everyone else's – just 50 per cent. There are many other examples of our tendency to fallacies in probabilistic reasoning, and many of them can – like this example – be partly explained by our tendency to seek and posit causal explanations where, seemingly, none are to be found.[15]

In light of this, here are some speculations about the real psychology of chance in humans, and how it compares with the formal model described above.

The formal model was completely general and somewhat unrealistic. Chance hypotheses emerged as a special case of Bayesian epistemology, where an agent has a certain sort of symmetry in her mental state: the distinctive attitude of treating trials as exchangeable. One way in which this formal model was obviously unrealistic was in the *non-dogmatism requirement*. This requirement, as used in the formal model, is completely implausible as an empirical claim. Finite organisms like us are not open to *every possible* hypothesis.

So what is going on in less cognitively sophisticated animals? I suggest that non-human animals have a much more limited 'hypothesis space' upon which their Bayesian updating works. Recall that the non-dogmatism requirement is the idea that you do not rule out any chance hypothesis as impossible. Clearly, some hypotheses are just too complicated for rats and pigeons to entertain. So these hypotheses *are*, in effect, ruled out a priori.

Relatively simple chance hypotheses – hypotheses that involve treating the data as exchangeable – are relatively *cheap* to include in one's hypothesis space, because (i) they are relatively easy to entertain, and (ii) they are

---

15 The claim that most humans are prone to fallacies in probabilistic reasoning is strongly associated with the research programme of Kahneman and Tversky (e.g. Tversky and Kahneman 1983 and many of the papers collected in Kahneman and Tversky 2000). Gerd Gigerenzer and David Murray (1987) have criticised this research programme, in part by denying that studies of this sort reveal that we are prone to fallacies of probabilistic reasoning in everyday contexts. Gigerenzer and Murray point out that in many of these laboratory settings it is reasonable for subjects to assume that there is a hidden pattern behind the phenomena, and that if they experienced such phenomena 'in the wild', they would not reason in the same way. For my purposes, I find this reply to have very little relevance to our overall assessment. We did not need lab studies to show that we are prone to superstitious theorising or other overconfident attribution of causes. A brief inspection of humans in their natural environment shows this to be extremely widespread.

For entertaining and popular discussion of our enthusiasm for causal narratives and consequent inability to recognise randomness, see Leonard Mlodinow's *The Drunkard's Walk* (2008) and Nassim Nicholas Taleb's *The Black Swan* (2010), especially chap. 6.

relatively easy to confirm and disconfirm. So something like simple chance hypotheses presumably play a big role in the probabilistic psychology of pigeons and rats. In contrast, humans have developed the ability to contemplate much more complicated hypotheses, many of which are not chance hypotheses, but are deterministic causal hypotheses, encompassing a huge range of variables. This is tremendously useful much of the time, because we so often identify what appear to be real causal relations between things that enable us to better manipulate the results for our desired ends. But we are so strongly inclined to engage in causal hypothesising that we sometimes posit causal relationships to explain merely random data. Hence we are prone to superstitious or ad hoc theorising.

Of course, I have no doubt that the cognitive psychology of humans differs from other animals in other ways too. But this difference seems to be potentially very important: our hypothesis space includes both chance hypotheses and deterministic causal hypotheses of relatively high complexity. The hypothesis space of a cognitively simpler organism is unlikely to include any but the simplest causal hypotheses, but might be expected to include many chance hypotheses.

Recall now the earlier worry that the Bayesian account fails to capture the 'physicality' of chance. When we say that 'quantum mechanics is chancy', we are – many people have thought – saying *more* than just 'we have found that using these Bayesian methods is more successful than using a deterministic causal hypothesis'. We are committed to some independent sort of chance fact, which cannot be reduced to frequencies, and which in turn *explains* why some frequencies are more probable than others. Although we have found reason to think that non-human animals may, like humans, employ something like Bayesian methods to confirm and disconfirm chance hypotheses, I take it that there is no reason to think that non-human animals are committed to these explicitly metaphysical or ontological claims about chance. Another way to put much the same point is that, while many human agents would expect it to be possible to offer an existential vindication of chance, in the form of substantive truth conditions for chance claims, it is not at all clear that any non-human animals have this expectation.

So there is something distinctive about our chance psychology that goes beyond the deployment of Bayesian techniques for handling uncertainty. Call this the 'ontological' aspect of our chance psychology. By this vague term I mean the urge to treat chance as an intrinsic property of experimental set-ups, as an independent matter of fact that plays an explanatory role much like causes, to postulate propensities, to seek truth conditions for chance ascriptions, and all manner of other philosophical projects in the broadly realist tradition.

Note that the entire project of justifying the principle of coordination (justifying the Principal Principle) only makes sense in light of an attitude which takes chances to be independent matters of fact about which we have beliefs, distinct from our credences. It would not make much sense to worry about a rat or a pigeon failing to ensure that its beliefs about the chances and its degrees of confidence coincide. For creatures like this, beliefs about chance *just are* degrees of confidence, combined with the distinctive attitude of exchangeability.

Having drawn attention to this distinctive ontological attitude to chance, the question to be asked from a natural history perspective is: what function does it serve? Why do we have it?

In light of the observation about our tendency to perform badly against rats and pigeons in chance set-ups, I suggest a hypothesis for the function of the ontological sophistication of chance. Part of the function of this concept is to *quell the urge* to seek further explanations. As we have already noted, sometimes it is maladaptive to seek further explanations. But having developed an aspect of our psychology that so persistently and doggedly seeks causal explanations and patterns in seemingly random phenomena, we have found it beneficial to posit an objective 'causal explanation cancelling' feature of the world: chance.

This hypothesis suggests an explanation of why we tend to treat chance as a physical property, with all the associated ontological baggage. Doing so makes chance more respectable as a way of explaining the world. If chances are merely subjectively optimal strategies for dealing with uncertainty, then it remains tempting to posit causal relationships that do better. But if chances are 'genuine physical properties', 'grounded in the intrinsic nature of the world', 'irreducible to anything else', and capable of 'genuinely explaining' what happens, then chances become a much more attractive place for our inquiring minds to come to rest.

Note that nothing I have said here particularly supports the thought that the ontological aspect of our chance concept is a trait that has evolved by natural selection. Indeed, this seems unlikely for two reasons. First, it seems unlikely that the concept is particularly *heritable* (though there are no doubt heritable cognitive capacities required to entertain the concept); and, second, the chance concept appears to have emerged much too quickly to be able to be explained in terms of natural selection.

Ian Hacking (1990) has written a history of the emergence of many of our ideas about chance, linking it to the development of statistical methods in the social sciences during the nineteenth century.

Hacking's account strongly suggests that the concept has been propagated overwhelmingly through cultural, rather than biological, means. As I mentioned in the case of morals, however, it is not essential to the debunking force of this explanation that the mechanism be by natural selection. All that is required is that the explanation we give does not presuppose the existence of chance, or the justification of the chance norms.

The vindication of chance?

In the case of moral concepts, having run into difficulty with the project of vindication, our turn to natural history supplied an *explanation* of why moral norms might exist, and why they might possess the peculiar character that distinguishes them from other norms. It ended up giving us a vindication, but a deflating one. Having put forward an admittedly sketchy and highly speculative account of the natural history of our chance psychology, are we better placed to address the vindication of chance?

We get the following sort of practical vindication of chance:

(4) For our ancestors (both human and non-human), for a reasonable variety of circumstances they encountered, it was adaptive to engage in probabilistic reasoning.

This seems very plausible, and by appeal to the sort of Bayesian account I sketched above, we can see how this might even get us far enough to vindicate ideas that serve a similar purpose as irreducibility and objectivity, as well as the chance norms. But it does not vindicate the distinctive ontological attitude we have to chance, such as the idea of physicality. If my speculation about the origin of that attitude is correct, then we can perhaps speculate that:

(5) For our human ancestors who had acquired the ability to engage in causal reasoning, it was, in some circumstances, adaptive to posit an ontologically loaded chance concept so as to discourage further seeking of causal explanations.

This has a nice parallel with the vindication given of moral judgments. Part of the role of moral judgments was to protect us from the potentially maladaptive effects of our overly flexible faculty for rationalisation. The role of an ontological concept of chance is to protect us from the maladaptive effects of our overeagerness to find causal hypotheses.

This is a rather attenuated practical vindication, but is it strictly deflating? I don't want to dwell on this issue. Certainly, it is better than nothing. Certainly, *I* do not plan on abandoning the chance norms.

Of more interest is the existential vindication. In light of this account, do we get a better sense of what it takes for a chance fact to obtain? I believe we get a strong presumption in favour of metaphysical anti-realism about chance. If chance psychology is useful simply when there is an *absence* of useful correlations that can be discerned by our faculty for causal reasoning, why think that this is evidence of some further chance fact? If my speculation about our chance psychology is correct – that the ontological nature of our chance concept has developed, not to track some independent feature of reality that we could just as well track without chance concepts, but to *discourage* us from using more 'high tech' cognitive apparatus in contexts where it will be fruitless – then this should greatly lower our confidence in the accuracy of this ontological aspect.

There is a parallel here with recent arguments for anti-realism about normativity due to Sharon Street (2006; forthcoming). Street, however, advocates a more thoroughgoing anti-realism about normativity than I am attempting to establish here. Her basic idea is that any realist account of normativity which supposes that we have evolved to track an independent normative fact is a less good explanation than an account which supposes that (i) we have merely evolved attitudes that have adaptive benefits, such as tracking biologically salient non-normative truths, and that (ii) the normativity is in some sense *constituted* by the attitudes themselves. If the normativity is constituted by our attitudes, then we have arrived at an anti-realist conclusion.

Analogously, in the current case, having become aware of a potential naturalistic explanation of why we might have the peculiar chance psychology that we do, and noting that this explanation does not require us to posit any tracking relation between objective chance facts and our chance psychology, we have reason to endorse an anti-realist account of chance as a superior explanation.[16]

---

16 If, despite these considerations, we insist on adopting some sort of realism about chance, it will have to be a *reductionist* realism. We would have to adopt a view about chance which abandons irreducibility – the claim that chances are irreducible to frequencies or other non-chancy facts. Any other view, which continues to assert irreducibility, will end up being vulnerable to a

In many philosophical discussions about chance, realism of some sort is assumed as a default position. In ethics, some variety of anti-realism is accorded much more respect. I claim that the two cases are less different than is usually supposed.

The case for anti-realism in ethics is especially strong, because it is so hard to vindicate the peculiarly strong force of moral norms. In comparison, practically vindicating the usefulness of the chance norms appears relatively easy. I do not claim that it has been done, but it does not look like a hopeless task.

When it comes to vindicating the *existence* of chance facts, however, things look extremely difficult. They are difficult in large part because chance is such a strangely irreducible notion. It defies simple identification with any pattern of empirically observable facts. And here there is a very strong parallel with morals, which similarly defy reduction to the naturalistic realm. Moreover, reflecting on the rather minimal psychology that is required to entertain probabilistic beliefs, and on the sort of vulnerabilities we have due to our ability to adopt causal hypotheses, lends further credence to the thought that our distinctive ontological ideas about chance might be useful, even if they do not track any mind-independent truth.[17]

Sometimes, when presenting this view to others, I meet with the reply:

> It is true that the nature of chance is mysterious. It is true that you can tell a naturalistic story in which the development of probabilistic beliefs is just a simple and cheap engineering fix by which we manage our limited powers to predict the future. And as far as that story goes, it undermines our confidence in the reality of chance. But *now that we have quantum mechanics* – a stunningly powerful theory by which we can predict events with great accuracy – and quantum mechanics tells us that the world is chancy, we have reason to think that we have discovered *real* chances.

As I have said, I don't despair of the practical utility of chance norms. In so far as quantum mechanics – or other successful scientific theories –

---

version of Street's argument. It will be entirely mysterious why the chance psychology which we have described above is going to evolve to reliably track the presence of the irreducible chance facts.

As already argued in Chapter 7, actualist accounts of chance – which must include all reductionist accounts of chance – fail to explain the normative role of chance unless they adopt an implausible anthropocentrism.

17 In the late stages of writing this book, I discovered a neglected short piece by R. F. Atkinson (1955), showing that A. J. Ayer's treatments of probability and ethics in *Language, Truth, and Logic* are inconsistent: that Ayer's arguments for anti-realism about ethical statements give rise to similar grounds for anti-realism about probability statements. My arguments are somewhat different from those considered by Ayer and Atkinson, but I am pleased to note the convergence upon a broadly similar result.

recommends that I adopt chancy beliefs, I am happy to do so. But regarding the metaphysical implications of these theories, I am less sanguine. Quantum mechanics does not provide any acceptable metaphysical account of the nature of chances. This goes largely unrecognised because there is very little understanding – among philosophers at least – of the content of quantum mechanics. As discussed in Chapter 9, some interpretations of quantum mechanics treat chance as primitive and unanalysed. Others treat it as analogous to the chances in statistical mechanics. Neither of these options provides an acceptable metaphysic of chance. The most tantalising option was the many-worlds interpretation. This promised a truly revolutionary understanding of chance as arising from a profound difference between the first-personal perspective of the world and the third-personal nature of the natural laws. But for the reasons discussed in Chapter 10, the many-worlds account of chance ultimately turns out to be another debunking account: a theory in which chance is not real, but in which it is not surprising that we believe in it.

So I recommend that we take an anti-realist attitude towards chance, even though chances are posited by our most successful physical theory. In doing so, I am at odds with scientific realists, who claim that we have good reason to think that successful theories are true – or nearly so. I reach this conclusion reluctantly, but no better opinion is available.

References

Abbott, Edwin A. 2007 [1884]. *Flatland: A Romance of Many Dimensions*. New York: Dover.

Albert, David Z. 1992. *Quantum Mechanics and Experience*. Cambridge, MA: Harvard University Press.

2000. *Time and Chance*. Cambridge, MA: Harvard University Press.

2010. 'Probability in the Everett Picture'. In Saunders *et al.* 2010: 355–68.

Allais, Maurice, and Ole Hagen, eds. 1979. *Expected Utility Hypotheses and the Allais Paradox*. Dordrecht: Reidel.

Arntzenius, Frank. 2000. 'Are There Really Instantaneous Velocities?' *The Monist* 83: 187–208.

Atkinson, R. F. 1955. '"Good" and "Right" and "Probable" in *Language, Truth, and Logic*'. *Mind* 64: 242–6.

Ayer, A. J. 1952. *Language, Truth, and Logic*. Revised edn. New York: Dover.

Barker, Stephen J. 2007. *Global Expressivism: Language Agency with Semantics, Reality without Metaphysics*. University of Nottingham ePrints.

Barrett, Jeffrey Alan. 1999. *The Quantum Mechanics of Minds and Worlds*. Oxford University Press.

Bass, Thomas A. 1985. *The Eudaemonic Pie*. Boston: Houghton Mifflin.

Bell, John. 1982. 'On the Impossible Pilot Wave'. *Foundations of Physics* 12: 989–99.

Bigelow, John. 1976. 'Possible Worlds Foundations for Probability'. *Journal of Philosophical Logic* 5: 299–320.

1977. 'Semantics of Probability'. *Synthese* 36: 459–72.

Bigelow, John, and Robert Pargetter. 1989. 'Vectors and Change'. *British Journal for the Philosophy of Science* 40: 289–306.

1990. *Science and Necessity*. Cambridge University Press.

Black, Max. 1954. 'The Identity of Indiscernibles'. In *Problems of Analysis*. London: Routledge and Kegan Paul, 80–92.

Black, Robert. 1998. 'Chance, Credence, and the Principal Principle'. *British Journal for the Philosophy of Science* 49: 371–85.

Blackburn, Simon. 1984. *Spreading the Word*. New York: Oxford University Press.

1993. *Essays in Quasi-Realism*. New York: Oxford University Press.

Bohm, David. 1952. 'A Suggested Interpretation of Quantum Theory in Terms of "Hidden Variables"'. *Physical Review* 85: 166–93.

Briggs, Rachael. 2009a. 'The Anatomy of the Big Bad Bug'. *Noûs* 43: 428–49.

2009b. 'The Big Bad Bug Bites Anti-realists about Chance'. *Synthese* 167: 81–92.

2009c. 'Distorted Reflection'. *Philosophical Review* 118: 59–85.

Broome, John. 2004. *Weighing Lives*. Oxford University Press.

Budescu, David V., and Thomas S. Wallsten. 1995. 'Processing Linguistic Probabilities: General Principles and Empirical Evidence'. *Psychology of Learning and Motivation* 32: 275–318.

Callender, Craig. 2004. 'There is No Puzzle about the Low Entropy Past'. In *Contemporary Debates in Philosophy of Science*, edited by Christopher Hitchcock. Oxford: Blackwell, chap. 12.

Carroll, John W. 1994. *Laws of Nature*. Cambridge University Press.

Chater, Nick, Joshua B. Tenenbaum, and Alan Yuille. 2006. 'Probabilistic Models of Cognition: Conceptual Foundations'. *Trends in Cognitive Sciences* 10: 287–91.

Christensen, David. 1991. 'Clever Bookies and Coherent Beliefs'. *Philosophical Review* 100: 229–47.

Collins, John, Ned Hall, and L. A. Paul, eds. 2004. *Causation and Counterfactuals*. Cambridge, MA: MIT Press.

de Finetti, Bruno. 1937. 'Foresight: Its Logical Laws, its Subjective Sources'. *Annales de l'Institut Henri Poincaré* 7. Reprinted and translated in Kyburg and Smokler 1964.

1974. *Theory of Probability*, translated by Antonio Machí and Adrian Smith. London: John Wiley & Sons.

DeRose, Keith. 1991. 'Epistemic Possibilities'. *Philosophical Review* 100: 581–605.

Deutsch, David. 1997. *The Fabric of Reality*. London: Penguin.

1999. 'Quantum Theory of Probability and Decisions'. *Proceedings of the Royal Society of London* A455: 3129–37.

DeWitt, Bryce S., and Neill Graham, eds. 1973. *The Many-Worlds Interpretation of Quantum Mechanics*. Princeton University Press.

Diaconis, P., and D. Freedman. 1980. 'Finite Exchangeable Sequences'. *Annals of Probability* 8: 745–64.

Dworkin, Ronald. 1977. *Taking Rights Seriously*. London: Duckworth.

1986. *Law's Empire*. Cambridge, MA: Harvard University Press.

Eagle, Antony. 2004. 'Twenty-one Arguments against Propensity Analyses of Probability'. *Erkenntnis* 60: 371–416.

ed. 2011. *Philosophy of Probability: Contemporary Readings*. London: Routledge.

Earman, John. 1986. *A Primer on Determinism*. Dordrecht: Reidel.

Eddington, A. S. 1929. *The Nature of the Physical World: Gifford Lectures, 1927*. New York: Macmillan.

Egan, Andy, John Hawthorne, and Brian Weatherson. 2005. 'Epistemic Modals in Context'. In *Contextualism in Philosophy: Knowledge, Meaning, and Truth*, edited by Gerhard Preyer and Georg Peter. Oxford University Press, 131–69.

Elga, Adam. 2001. 'Statistical Mechanics and the Asymmetry of Counterfactual Dependence'. *Philosophy of Science* 68: 313–24.

2004. 'Infinitesimal Chances and the Laws of Nature'. *Australasian Journal of Philosophy* 82: 67–76.

Eriksson, Lina, and Alan Hájek. 2007. 'What Are Degrees of Belief?' *Studia Logica* 86: 183–213.

Everett, Hugh, III. 1957a. 'On the Foundations of Quantum Mechanics'. Ph.D. thesis, Princeton University.

1957b. ' "Relative State" Formulation of Quantum Mechanics'. *Reviews of Modern Physics* 29: 454–62. Reprinted in Wheeler and Zurek 1983.

Foot, Philippa. 1972. 'Morality as a System of Hypothetical Imperatives'. *Philosophical Review* 81: 305–16.

Frigg, Roman. 2011. 'Why Typicality Does Not Explain the Approach to Equilibrium'. In *Probabilities, Causes, and Propensities in Physics*, edited by Mauricio Suárez. Dordrecht: Springer, 77–93.

Frigg, Roman, and Carl Hoefer. 2007. 'Probability in GRW Theory'. *Studies in History and Philosophy of Modern Physics* 38: 371–89.

Frisch, Mathias. 2005. 'Counterfactuals and the Past Hypothesis'. *Philosophy of Science* 72: 739–50.

2007. 'Causation, Counterfactuals, and Entropy'. In *Causation, Physics, and the Constitution of Reality*, edited by Huw Price and Richard Corry. Oxford University Press, 351–95.

2010. 'Does a Low-Entropy Constraint Prevent Us from Influencing the Past?' In *Time, Chance, and Reduction: Philosophical Aspects of Statistical Mechanics*, edited by Andreas Hüttemann and Gerhard Ernst. Cambridge University Press, 13–33.

Galavotti, Maria Carla. 2005. *A Philosophical Introduction to Probability*. Stanford, CA: CSLI Publications.

Ghirardi, G. C., A. Rimini, and T. Weber. 1986. 'Unified Dynamics for Microscopic and Macroscopic Systems'. *Physical Review D* 34: 470–91.

Gigerenzer, Gerd, and David J. Murray. 1987. *Cognition as Intuitive Statistics*. Hillsdale, NJ: Erlbaum.

Greaves, Hilary. 2007. 'Probability in the Everett Interpretation'. *Philosophy Compass* 2: 109–28.

Greaves, Hilary, and Wayne C. Myrvold. 2010. 'Everett and Evidence'. In Saunders *et al.* 2010: 264–304.

Hacking, Ian. 1967. 'Possibility'. *Philosophical Review* 76: 143–68.

1990. *The Taming of Chance*. Cambridge University Press.

Haidt, Jonathan, and Craig Joseph. 2004. 'Intuitive Ethics: How Innately Prepared Intuitions Generate Culturally Variable Virtues'. *Daedalus* 133: 55–66.

Hájek, Alan. 1997. ' "Mises Redux" – Redux: Fifteen Arguments against Finite Frequentism'. *Erkenntnis* 45: 209–27.

2003a. 'Conditional Probability Is the Very Guide of Life'. In *Probability Is the Very Guide of Life: The Philosophical Uses of Chance*, edited by Henry E. Kyburg, Jr and Mariam Thalos. Chicago: Open Court, 183–203.

2003b. 'What Conditional Probability Could Not Be'. *Synthese* 137: 273–323.

2005. 'Scotching Dutch Books'. *Philosophical Perspectives* 19: 139–51.

2010. 'Interpretations of Probability'. In *The Stanford Encyclopedia of Philosophy*, edited by Edward N. Zalta, Spring 2010 edn. http://plato.stanford.edu/archives/spr2010/entries/probability-interpret/.

Hall, Ned. 1994. 'Correcting the Guide to Objective Chance'. *Mind* 103: 505–17.

2004. 'Two Mistakes about Credence and Chance'. *Australasian Journal of Philosophy* 82: 93–111.

Hare, Caspar. 2009. *On Myself, and Other, Less Important Subjects*. Princeton University Press.

Hart, H. L. A. 1961. *The Concept of Law*. Oxford: Clarendon Press.

Hawthorne, John, and Maria Lasonen-Aarnio. 2009. 'Knowledge and Objective Chance'. In *Williamson on Knowledge*, edited by Patrick Greenough and Duncan Pritchard. Oxford University Press, 92–108.

Hoefer, Carl. 1997. 'On Lewis's Objective Chance: "Humean Supervenience Debugged"'. *Mind* 106: 321–34.

2007. 'The Third Way on Objective Probability: A Sceptic's Guide to Objective Chance'. *Mind* 116: 549–96.

Holton, Richard. 2008. 'Partial Belief, Partial Intention'. *Mind* 117: 27–58.

2009. *Willing, Wanting, Waiting*. Oxford: Clarendon Press.

Horwich, Paul. 1987. *Asymmetries in Time*. Cambridge, MA: MIT Press.

Howson, Colin, and Peter Urbach. 1989. *Scientific Reasoning: The Bayesian Approach*. First edn. Chicago: Open Court.

Hume, David. 1902 [1777]. *An Enquiry Concerning Human Understanding*, edited by M. A. Selby-Bigge. Second edn. Oxford: Clarendon Press.

Ismael, Jenann. 2009. 'Probability in Deterministic Physics'. *Journal of Philosophy* 106: 89–108.

Jackson, Frank, and Philip Pettit. 1998. 'A Problem for Expressivism'. *Analysis* 58: 239–51.

Joyce, James M. 2007. 'Epistemic Deference: The Case of Chance'. *Proceedings of the Aristotelian Society* 107: 187–206.

Joyce, Richard. 2006. *The Evolution of Morality*. Cambridge, MA: MIT Press.

Kahneman, Daniel, and Amos Tversky, eds. 2000. *Choices, Values, and Frames*. Cambridge University Press.

Kemeny, John G. 1955. 'Fair Bets and Inductive Probabilities'. *Journal of Symbolic Logic* 20: 263–73.

Kolmogorov, A. N. 1956. *Foundations of the Theory of Probability*, translation edited by Nathan Morrison. Second edn. New York: Chelsea Publishing Company.

Kratzer, Angelika. 1977. 'What "Must" and "Can" Must and Can Mean'. *Linguistics and Philosophy* 1: 337–55.

1991. 'Modality'. In *Semantics: An International Handbook of Contemporary Research*, edited by Arnim von Stechow and Dieter Wunderlich. Berlin: De Gruyter, 639–50.

Kyburg, Henry E., Jr. 1978. 'Subjective Probability: Criticisms, Reflections, and Problems'. *Journal of Philosophical Logic* 7: 157–80. Reprinted in Eagle 2011.

Kyburg, Henry E., Jr, and Howard E. Smokler, eds. 1964. *Studies in Subjective Probability*. New York: John Wiley & Sons.

Langer, Ellen J., and Jane Roth. 1975. 'Heads I Win, Tails it's Chance: The Illusion of Control as a Function of the Sequence of Outcomes in a Purely Chance Task'. *Journal of Personality and Social Psychology* 32: 951–5.

Lehman, R. Sherman. 1955. 'On Confirmation and Rational Betting'. *Journal of Symbolic Logic* 20: 251–62.

Lewis, David. 1973a. 'Causation'. *Journal of Philosophy* 70: 556–67. Reprinted (plus postscripts) in Lewis 1986b: 159–213.

   1973b. *Counterfactuals*. Oxford: Blackwell.

   1976a. 'The Paradoxes of Time Travel'. *American Philosophical Quarterly* 13: 145–52.

   1976b. 'Survival and Identity'. In *The Identities of Persons*, edited by Amelie O. Rorty. Berkeley: University of California Press, 17–40. Reprinted in Lewis 1983b: 55–77.

   1979a. 'Attitudes De Dicto and De Se'. *Philosophical Review* 88: 513–43.

   1979b. 'Counterfactual Dependence and Time's Arrow'. *Noûs* 13: 455–76. Reprinted (plus postscripts) in Lewis 1986b: 32–65.

   1980. 'A Subjectivist's Guide to Objective Chance'. In *Studies in Inductive Logic and Probability*, edited by Richard C. Jeffrey, vol. II. Berkeley: University of California Press. Reprinted in Lewis 1986b.

   1983a. 'New Work for a Theory of Universals'. *Australasian Journal of Philosophy* 61: 343–77. Reprinted in Lewis 1999: 8–55.

   1983b. *Philosophical Papers*, vol. I. New York: Oxford University Press.

   1986a. *On the Plurality of Worlds*. Oxford: Blackwell.

   1986b. *Philosophical Papers*, vol. II. New York: Oxford University Press.

   1991. *Parts of Classes*. Oxford: Blackwell.

   1994. 'Humean Supervenience Debugged'. *Mind* 103: 473–90. Reprinted in Lewis 1999: 224–47.

   1999. *Papers in Metaphysics and Epistemology*. Cambridge University Press.

   2004. 'How Many Lives Has Schrödinger's Cat?' *Australasian Journal of Philosophy* 82: 3–22.

Lewis, Peter J. 2010. 'Probability in Everettian Quantum Mechanics'. *Manuscrito* 33: 285–306.

Loewer, Barry. 2004. 'David Lewis's Humean Theory of Objective Chance'. *Philosophy of Science* 71: 1115–25.

   2007. 'Counterfactuals and the Second Law'. In *Causation, Physics, and the Constitution of Reality*, edited by Huw Price and Richard Corry. Oxford University Press, 293–326.

   Forthcoming. 'Two Accounts of Laws and Time'. *Philosophical Studies*.

Logue, James. 1995. *Projective Probability*. Oxford University Press.

Lyon, Aidan. 2010. 'Deterministic Probability: Neither Chance nor Credence'. *Synthese* 182: 413–32.

MacFarlane, John. 2005. 'The Assessment Sensitivity of Knowledge Attributions'. *Oxford Studies in Epistemology* 1: 197–233.

Maudlin, Tim. 1995. 'Three Measurement Problems'. *Topoi* 14: 7–15.

2007a. *The Metaphysics within Physics*. Oxford University Press.

2007b. 'What Could Be Objective about Probabilities?' *Studies in History and Philosophy of Modern Physics* 38: 275–91.

McCall, Storrs. 1994. *A Model of the Universe*. Oxford: Clarendon Press.

Mellor, D. H. 1971. *The Matter of Chance*. Cambridge University Press.

Mlodinow, Leonard. 2008. *The Drunkard's Walk*. New York: Pantheon.

Moore, G. E. 1993. *Principia Ethica*. Revised edn. Cambridge University Press.

Ninan, Dilip. 2009. 'Persistence and the First-Person Perspective'. *Philosophical Review* 118: 425–64.

Nolan, Daniel. 2006. 'Stoic Gunk'. *Phronesis* 51: 162–83.

Norton, John D. 2008. 'The Dome: An Unexpected Simple Failure of Determinism'. *Philosophy of Science* 75: 786–98.

Parfit, Derek. 1985. *Reasons and Persons*. Oxford University Press.

Perry, John. 1977. 'Frege on Demonstratives'. *Philosophical Review* 86: 474–97.

1979. 'The Problem of the Essential Indexical'. *Noûs* 13: 3–21.

Popper, Karl R. 1957. 'The Propensity Interpretation of the Calculus of Probability, and the Quantum Theory'. In *Observation and Interpretation*, edited by Stephan Körner. London: Butterworths, 65–70. Reprinted in Sklar 2000.

Price, Huw. 1996. *Time's Arrow and Archimedes' Point*. Oxford University Press.

2004. 'On the Origins of the Arrow of Time: Why There Is Still a Puzzle about the Low Entropy Past'. In *Contemporary Debates in Philosophy of Science*, edited by Christopher Hitchcock. Oxford: Blackwell, chap. 11.

2010. 'Decisions, Decisions, Decisions: Can Savage Salvage Everettian Probability?' In Saunders *et al.* 2010: 369–90.

2011. 'The Flow of Time'. In *The Oxford Handbook of Philosophy of Time*, edited by Craig Callender. Oxford University Press, chap. 9.

Price, Huw, and Brad Weslake. 2010. 'The Time-Asymmetry of Causation'. In *The Oxford Handbook of Causation*, edited by Helen Beebee, Christopher Hitchcock, and Peter Menzies. Oxford University Press, 414–43.

Ramsey, F. P. 1931. 'Truth and Probability'. In *The Foundations of Mathematics and Other Logical Essays*. London: Routledge and Kegan Paul, 156–98.

Raz, Joseph. 1985. 'Value Incommensurability: Some Preliminaries'. *Proceedings of the Aristotelian Society* 86: 117–34.

Saunders, Simon. 1998. 'Time, Quantum Mechanics, and Probability'. *Synthese* 114: 373–404.

2010. 'Chance in the Everett Interpretation'. In Saunders *et al.* 2010: 181–205.

Saunders, Simon, Jonathan Barrett, Adrian Kent, and David Wallace, eds. 2010. *Many Worlds? Everett, Quantum Theory, and Reality*. Oxford University Press.

Saunders, Simon, and David Wallace. 2008. 'Branching and Uncertainty'. *British Journal for the Philosophy of Science* 59: 293–305.

Savage, Leonard J. 1954. *The Foundations of Statistics*. New York: John Wiley & Sons.

Schaffer, Jonathan. 2003a. 'Is There a Fundamental Level?' *Noûs* 37: 498–517.

2003b. 'Principled Chances'. *British Journal for the Philosophy of Science* 54: 27–41.

2007. 'Deterministic Chance?' *British Journal for the Philosophy of Science* 58: 113–40.

Schroeder, Mark. 2008. *Being For: Evaluating the Semantic Program of Expressivism*. Oxford: Clarendon Press.

Sklar, Lawrence, ed. 2000. *Probability and Confirmation*, vol. IV of *The Philosophy of Science: A Collection of Essays*. New York: Garland.

Skyrms, Brian. 1984. *Pragmatics and Empiricism*. New Haven, CT: Yale University Press.

Sorensen, Roy. 2008. 'Vagueness'. In *The Stanford Encyclopedia of Philosophy*, edited by Edward N. Zalta, Fall 2008 edn. http://plato.stanford.edu/archives/fall2008/entries/vagueness/.

Strawson, Galen. 1992. *The Secret Connexion: Causation, Realism, and David Hume*. New York: Oxford University Press.

Street, Sharon. 2006. 'A Darwinian Dilemma for Realist Theories of Value'. *Philosophical Studies* 127: 109–66.

Forthcoming. 'Evolution and the Normativity of Epistemic Reasons'. In *Belief and Agency*, edited by David Hunter. *Canadian Journal of Philosophy*, Supplemental Volume, Calgary: University of Calgary Press.

Strevens, Michael. 1995. 'A Closer Look at the "New" Principle'. *British Journal for the Philosophy of Science* 46: 545–61.

1998. 'Inferring Probabilities from Symmetries'. *Noûs* 32: 231–46.

2003. *Bigger than Chaos: Understanding Complexity through Probability*. Cambridge, MA: Harvard University Press.

2011. 'Probability out of Determinism'. In *Probabilities in Physics*, edited by Claus Beisbart and Stephan Hartmann. Oxford University Press.

Taleb, Nassim Nicholas. 2010. *The Black Swan: The Impact of the Highly Improbable*. Revised edn. London: Penguin.

Teller, Paul. 1973. 'Conditionalization and Observation'. *Synthese* 26: 218–58.

Thau, Michael. 1994. 'Undermining and Admissibility'. *Mind* 103: 491–503.

Tooley, Michael. 1988. 'In Defense of the Existence of States of Motion'. *Philosophical Topics* 16: 225–54.

Tversky, Amos, and Daniel Kahneman. 1983. 'Extensional versus Intuitive Reasoning: The Conjunction Fallacy in Probability Judgment'. *Psychological Review* 90: 293–315.

van Fraassen, Bas C. 1989. *Laws and Symmetry*. Oxford: Clarendon Press.

van Roojen, Mark. 2011. 'Moral Cognitivism vs. Non-Cognitivism'. In *The Stanford Encyclopedia of Philosophy*, edited by Edward N. Zalta, Spring 2011 edn. http://plato.stanford.edu/archives/spr2011/entries/moral-cognitivism/.

von Fintel, Kai, and Anthony S. Gillies. 2008. 'CIA Leaks'. *Philosophical Review* 117: 77–98.

Wallace, David. 2002. 'Quantum Probability and Decision Theory, Revisited'. arXiv. org:quant-ph/0211104.

2003. 'Everett and Structure'. *Studies in History and Philosophy of Modern Physics* 34: 87–105.

2007. 'Quantum Probability from Subjective Likelihood: Improving on Deutsch's Proof of the Probability Rule'. *Studies in History and Philosophy of Modern Physics* 38: 311–32.

Forthcoming. *The Emergent Multiverse.* Oxford: Clarendon Press.

Ward, Barry. 2005. 'Projecting Chances: A Humean Vindication and Justification of the Principal Principle'. *Philosophy of Science* 72: 241–61.

Wheeler, John Archibald, and Wojciech Herbert Zurek, eds. 1983. *Quantum Theory and Measurement.* Princeton University Press.

White, Roger. 2005. 'Epistemic Permissiveness'. *Philosophical Perspectives* 19: 445–59.

2010. 'Evidential Symmetry and Mushy Credence'. *Oxford Studies in Epistemology* 3: 161–86.

Williams, Bernard. 1970. 'The Self and the Future'. *Philosophical Review* 79: 161–80.

Williams, J. Robert G. 2008. 'Chances, Counterfactuals, and Similarity'. *Philosophy and Phenomenological Research* 77: 385–420.

Wilson, Alastair. 2011. 'Macroscopic Ontology in Everettian Quantum Mechanics'. *Philosophical Quarterly* 61: 362–82.

Wolford, George, Michael B. Miller, and Michael Gazzaniga. 2000. 'The Left Hemisphere's Role in Hypothesis Formation'. *Journal of Neuroscience* RC 64: 1–4.

Yalcin, Seth. 2007. 'Epistemic Modals'. *Mind* 116: 983–1026.

Yellott, John I. 1969. 'Probability Learning with Noncontingent Success'. *Journal of Mathematical Psychology* 6: 541–75.

Index

actualism, *see* chance, actualism regarding
Albert, David, 206, 208–13
anti-realism, *see also* realism
    regarding chance, 123–45, 158, 191, 243–5
    regarding morals, 226, 244
Ayer, A. J., 22

Bayesian epistemology, 137, 140–1, 182–7, 232–7,
    see also degrees of belief, updating by
    conditionalisation
    evolution and, 190
Bernoulli process, 138
best-system account of laws and chance, 110–13,
    143, 157
    fit in, 114–16, 119, 120
    simplicity in, 111–12, 118–19, 121
    strength in, 112–13, 121
Bigelow, John, 78, *see also* chance, possibilism
    regarding
Blackburn, Simon, 143
Bohmian mechanics, 146, 155–8
branching worlds, 159–62, 174, *see also*
    Everettian quantum mechanics
Briggs, Rachael, 143

causation, 104, 199
chance, *see also* anti-realism regarding chance;
    contextual variability of chance
    ascriptions
    actualism regarding, 31, 104–22, 229, *see also*
    chance, frequentist accounts of
    anthropocentrism regarding, 121–2
    as a subspecies of probability, 127
    as an advice function, 16–22, 71, 89, 95, 216
    conditional on measure zero propositions,
    79–81
    frequentist accounts of, 106–9, 119, 121
    in scientific explanations, 3, 88, 95, 101, 118,
    126, 130, 158, 187
    normative role of, 4, 22–3, 88, 89, 92–5, 103,
    120–1, 127, 187, 227–31, 241–3
    objectivity of, 26–8, 88, 101, 106, 123, 131, 158,
    228, 235

    of past events, 7–8, 29, *see also* temporal
    asymmetry, of chances
    physicality of, 2–4, 85–6, 229, 241
    possibilism regarding, 31, 78–103
    primitivism regarding, 32
    propensity accounts of, 96
    subjectivist accounts of, 123, 127–31, 142
classical physics, 34–46, *see also* determinism;
    statistical mechanics
    gases in, 69–70
    particles in, 35–6, 41
    velocity in, 39, 44, 59
conditionalisation, *see* Bayesian epistemology
contextual variability of chance ascriptions, 24,
    123, 216–17
counterfactuals, 67–8, 117
credences, *see* degrees of belief
Curie, Marie, 26

de Finetti, Bruno, 129, 234, *see also* degrees of
    belief, exchangeable
decision theory, 132–4
    Dominance axiom of, 180
degrees of belief, 2, 8–14
    as a species of probability, 12–14, 127
    characterised by way of betting dispositions,
    9–11, 132, 135–6, 177, 184–6
    distinct from ordinary belief, 9
    exchangeable, 137–40, 184, 233–5, 239
    rational constraints upon, 101, 130, 177–82,
    187, 189, *see also* chance, normative role
    of
    regarding initial conditions, 83, 96–8, 103,
    158
    scientific evidence for, 11
    updating by conditionalisation, 130, 137, 233,
    see also Bayesian epistemology
determinism, 39, 42, 61, 68, 97, 103
    and quantum mechanics, 158, *see also*
    Bohmian mechanics
    compatibility of chance and, 29, 30, 73, 83–5,
    172
Deutsch, David, 160, 176, 183

254

DeWitt, Bryce, 159
Dutch Book arguments, 12–14
Dworkin, Ronald, 222

Eagle, Antony, 89
Elga, Adam, 115
entropy, *see* thermodynamics, second
    law of
epistemic modals, 24
error theory, 124–7, 191, *see also* anti-realism
    regarding chance
Everettian quantum mechanics, 156,
    159–61
    and branch weight, 182, 186–8
    and posthumous events, 172–6, 186
    personal identity in, 162–6, 170
    uncertainty in, 162–76, 184
evidence
    admissibility of, *see* Principal Principle, the
    availability of, 23–9, 71, 81, 84, 215
    for chances, *see* frequencies; symmetry,
      physical

Foot, Philippa, 220
frequencies, 31, 229
Frigg, Roman, 157
fundamentality, 36, 65

gambling, 1
    apparatus, 2, 25
    aversion to, 10
Gigerenzer, Gerd, 239
Greaves, Hilary, 182

Hacking, Ian, 24
Hart, H. L. A., 221
heat, *see* thermodynamics
Hoefer, Carl, 121, 157
Holton, Richard, 9
Hume, David, 104, 135
Humeanism, 31, 104–5, 110
Hájek, Alan, 80, *see also* reference class
    problem

ignorance, 83, 88, 101, *see also* probability,
    epistemic
incommensurate value, 134
intrinsicness, *see* chance, physicality of
invariance under time-reversal, *see*
    time-reversibility in physics
Ismael, Jenann, 103

Joyce, Richard, 224

Kahneman, Daniel, 91, 239
Kolmogorov, A. N., *see* probability, axioms of
Kyburg, Henry E., Jr, 13

Laplace, Pierre-Simon, 89
laws of nature, 30, 86, 110–11, 172, *see also*
    best-system account of laws and chance
    deterministic, *see* determinism
    gravitational, 55, 65
Lewis, David, 18, 31, 89, 105, 169–70, 190, 197,
    229, *see also* best-system account of laws
    and chance; Principal Principle, the
Loewer, Barry, 213
Logue, James, 143
Lyon, Aidan, 97

Markov process, 138
matter, divisibility of, 35, 65
Maudlin, Tim, 32, 97, 105, 158, 200–1
McCall, Storrs, 160
measure, mathematical, 75, 90, 96, 102, 158,
    176
    Lebesgue, 76
measurement, 208–11
method of arbitrary functions, *see* probability,
    microconstancy account of
Murray, David, 239
Myrvold, Wayne, 182

Newtonian mechanics, *see* classical physics
Ninan, Dilip, 171
non-cognitivism, 124, 142–5
normativity, *see also* chance, normative role of
    legal, 219–22
    moral, 219–20

phase space, 56–60, 65, 67, 72, 87
possibilism, *see* chance, possibilism regarding
possibilities, 43, 45, 47–52, 92, *see also* phase
    space
    counterlegal, 53–5, 65–6, 81
    egocentric, 166–9, 172
    infinite number of, 52, 75, 90
    mechanical, 55–6
    possible histories, 59–61, 85
possible worlds, *see* possibilities, possible
    histories
Price, Huw, 217
Principal Principle, the, 18, 227
    and admissibility of evidence, 18–21,
      29
    justification of, 120, 122, 241, *see also* chance,
      normative role of

principle of indifference, 101
probability
    axioms of, 12, 14–15, 80, 91, 129
    classical account of, 89–90
    distinguished from chance, *see* chance, as a
        subspecies of probability
    epistemic, 28, 83, 88, 95, 157
    microconstancy account of, 97–103,
        158–9
propositions, *see also* possibilities
    as sets of possibilities, 62–71
    impossible, 64–6
    self-locating, 169–71, *see also* possibilities,
        egocentric
    versus sentences, as arguments of chance,
        5–7

quantum mechanics, 126, *see also* Bohmian
        mechanics
    collapse of the wave function, 154–5
    Ghirardi, Rimini, and Weber (GRW)
        interpretation, 155, 157
    many-worlds interpretation of, *see*
        Everettian quantum mechanics
    non-locality of, 150, 160
    particles in, 147–50
    superpositions, 152–4, 161

radioactive decay, 192–3
Ramsey, F. P., 9, 10, 14
rationality, *see* chance, normative role of;
        degrees of belief, rational constraints
        upon
realism, *see also* anti-realism
    regarding chance, 121, 124
    scientific, 126, 244–5
reference class problem, 106–7

Saunders, Simon, 171, 174, 186
Savage, Leonard J., 11, 135

Schaffer, Jonathan, 29, 83–7, *see also*
        fundamentality
Skyrms, Brian, 131
space, absolutism versus relationalism, 50–1
statistical mechanics, 69–70, 76–8, 83, 89, *see*
        also thermodynamics
statistical postulate, 77, 207, 210
Strawson, Galen, 104
Strevens, Michael, 97, 158, *see also* probability,
        microconstancy account of
supervaluation, 82
symmetry, physical, 180

temporal asymmetry, 30, 46, *see also*
        thermodynamics, second law of;
        time-reversibility in physics
    of causation, 198–9, 213
    of chances, 86, 192–5
    of evidence, 195–7, 208–14
thermodynamics, 201–3
    and the past hypothesis, 206–14
    second law of, 73, 74, 203–7, 211
time-reversibility in physics, 43–6, 48, 72–3,
        199–201, 204–6
time travel, 8
Tversky, Amos, 91, 239

utility, *see* decision theory; degrees of belief,
        characterised by way of betting
        dispositions

vagueness, 80, 82
volume, *see* measure, mathematical

Wallace, David, 171, 177, 183, 186
Ward, Barry, 143
ways the world could be, *see* possibilities
weakness of will, 224
White, Roger, 101
Wilson, Alastair, 173–4